SYMPOSIA OF THE
SOCIETY FOR EXPERIMENTAL BIOLOGY

NUMBER L

SYMPOSIA OF THE
SOCIETY FOR EXPERIMENTAL BIOLOGY

I	Nucleic Acid
II	Growth, Differentiation and Morphogenesis
III	Selective Toxicity and Antibiotics
IV	Physiological Mechanisms in Animal Behaviour
V	Fixation of Carbon Dioxide
VI	Structural Aspects of Cell Physiology
VII	Evolution
VIII	Active Transport and Secretion
IX	Fibrous Proteins and their Biological Significance
X	Mitochondria and other Cytoplasmic Inclusions
XI	Biological Action of Growth Substances
XII	The Biological Replication of Macromolecules
XIII	Utilization of Nitrogen and its Compounds by Plants
XIV	Models and Analogues in Biology
XV	Mechanisms in Biological Competition
XVI	Biological Receptor Mechanisms
XVII	Cell Differentiation
XVIII	Homeostasis and Feedback Mechanisms
XIX	The State and Movement of Water in Living Organisms
XX	Nervous and Hormonal Mechanisms of Integration
XXI	Aspects of the Biology of Ageing
XXII	Aspects of Cell Motility
XXIII	Dormancy and Survival
XXIV	Control of Organelle Development
XXV	Control Mechanisms of Growth and Differentiation
XXVI	The Effects of Pressure on Organisms
XXVII	Rate Control of Biological Processes
XXVIII	Transport at the Cellular Level
XXIX	Symbiosis
XXX	Calcium in Biological Systems
XXXI	Integration of Activity in the Higher Plant
XXXII	Cell–Cell Recognition
XXXIII	Secretory Mechanisms
XXXIV	The Mechanical Properties of Biological Materials
XXXV	Prokaryotic and Eukaryotic Flagella
XXXVI	The Biology of Photoreceptors
XXXVII	Neural Origin of Rhythmic Movements
XXXVIII	Controlling Events in Meiosis
XXXIX	Physiological Adaptations of Marine Animals
XL	Plasticity in Plants
XLI	Temperature and Animals Cells
XLII	Plants and Temperature
XLIII	Mucus and Related Topics
XLIV	Hormone Perception and Signal Transduction in Animals and Plants
XLV	Molecular Biology of Plant Development
XLVI	Molecular Biology of Muscle
XLVII	Cell Behaviour: Adhesion and Motility
XLVIII	Membrane Transport in Plants and Fungi: Molecular Mechanisms and Control
XLIX	Biological Fluid Dynamics

SYMPOSIA OF THE
SOCIETY FOR EXPERIMENTAL BIOLOGY

NUMBER L

UNIFYING PLANT GENOMES

EDITED BY
J. S. HESLOP-HARRISON

SOCIETY FOR EXPERIMENTAL BIOLOGY SYMPOSIA

SEB Symposia form a long-standing series of volumes first published in 1947. The series is annual and each publication is a collection of authoritative articles on an aspect of modern experimental biology.

The aims of the Symposium series are to stimulate discussion and communication between scientists of all nationalities, and to foster the development of, and research on, modern aspects of plant, animal and cell biology.

Published for the Society for Experimental Biology
by The Company of Biologists Limited
Bidder Building, 140 Cowley Road, Cambridge CB4 4DL, UK

Typeset, Printed and Published by The Company of Biologists Limited,
Bidder Building, 140 Cowley Road, Cambridge CB4 4DL, UK

© Society for Experimental Biology 1996
ISBN 0 948601 51 5
ISSN 0081–1386

Cover picture
Fluorescence in situ hybridization to root tip metaphase chromosomes of a wheat-alien line. Total genomic DNA from rye identifies the two rye chromosome arms (green fluorescence) and DNA from a wild couch grass (*Thynopyron bessarabicum*) identifies two added chromosomes (red fluorescence). The 42 wheat chromosomes fluoresce blue with the DNA stain DAPI (see article by T. Schwarzacher, pp. 71-75). The fluorescence micrograph is overlaid by comparative maps of a segment of the wheat, sorghum and maize genomes, showing a paracentric inversion specific to maize. The filled circles indicate known centromere positions, and the open circle indicates the predicted centromere position in sorghum. These maps were drawn by Dr Jeff Bennetzen using data from his laboratory and the laboratory of Dr Mike Gale.

CONTENTS

Preface

THE HYBAID LECTURE
Bennetzen, J. L., SanMiguel, P., Liu, C.-N., Chen, M., Tikhonov, A., Costa de Oliveira, A., Jin, Y.-K., Avramova, Z., Woo, S.-S., Zhang, H. and Wing, R. A.
Microcollinearity and segmental duplication in the evolution of grass nuclear genomes — 1-3

Delseny, M., Raynal, M., Laudié, M., Varoquaux, F., Comella, P., Wu, H. J., Cooke, R. and Grellet, F.
Sequencing and mapping the *Arabidopsis* genome: a weed model for real crops — 5-9

Sasaki, T.
Rice cDNAs as a model for expressed genes of plants — 11-15

Heslop-Harrison, J. S.
Comparative analysis of plant genome architecture — 17-23

Herrmann, R. G., Martin, R., Busch, W., Wanner, G. and Hohmann, U.
Physical and topographical mapping among Triticeae chromosomes — 25-30

Lee, M.
Comparative genetic and QTL mapping in sorghum and maize — 31-38

Hoisington, D., Jiang, C., Khairallah, M., Ribaut, J.-M., Bohn, M., Melchinger, A., Willcox, M. and González-de-León, D.
QTL for insect resistance and drought tolerance in tropical maize: prospects for marker assisted selection — 39-44

Bennett, M. D.
The nucleotype, the natural karyotype and the ancestral genome — 45-52

Doudrick, R. L.
Genetic recombinational and physical linkage analyses on slash pine — 53-60

Jacobsen, N. and Orgaard, M.
Unifying plant molecular data and plants — 61-64

Motoyoshi, F., Ohmori, T. and Murata, M.
Molecular characterization of heterochromatic regions around the *Tm-2* locus in chromosome 9 of tomato — 65-70

Schwarzacher, T.
The physical organization of Triticeae chromosomes — 71-75

Index — 77-78

PREFACE

There is an enormous amount of data about the genomes of individual plant species, but only now can we begin to compare the genomes of different species. The authors of the chapters in this volume highlight the common elements of genome organization which are present in genera, families or even kingdoms, and explore aspects of the higher-order structure, evolution, variation and organization of the whole plant genome and its components. Scientists contributing to the book include those involved in the major genome projects, such as those for rice and *Arabidopsis*, as well as species of practical importance such as timber pines and wheat. Such species, with huge genomes, were often regarded as intractable to molecular genetic analysis but the combination of new methods and background knowledge of plant genome organization is advancing genome studies even in these species. This volume marks the start of a new era of plant genome studies involving the integration of genetical, physical and structural models of plant genomes. The models arising from such works will be valuable in the evaluation and utilization of biodiversity and in plant breeding, and we hope that the volume will prove to be useful to both research scientists and students involved in understanding plant genomes.

J. S. H.-H.

Printed in Great Britain © The Society for Experimental Biology 1996
SEB0022

THE HYBAID LECTURE

Microcollinearity and segmental duplication in the evolution of grass nuclear genomes

Jeffrey L. Bennetzen[1], Phillip SanMiguel[1], Chang-Nong Liu[1], Mingsheng Chen[1], Alexander Tikhonov[1], Antonio Costa de Oliveira[1], Young-Kwan Jin[1], Zoya Avramova[1], Sung-Sick Woo[2], Hongbin Zhang[2] and Rod A. Wing[2]

Department of Biological Sciences, Purdue University, West Lafayette, IN 47907-1392, USA
Department of Soil and Crop Sciences, Texas A&M University, College Station, TX 77843-2123, USA

SUMMARY

Recent studies have shown that grass genomes have very similar gene compositions and regions of conserved gene order, as exemplified by collinear genetic maps of DNA markers. We have begun the detailed study of sequence organization in large (100-500 kb) segments of the nuclear genomes of maize, sorghum and rice. Our results indicate collinearity of genes in the regions homoeologous to the maize *adh1* and *sh2-a1* genes. Comparable genes were found to be physically closer to each other in grasses with small genomes (rice and sorghum) than they are in maize.

In several instances, we have found evidence of tandem and 'distantly tandem' duplications of segments containing maize and sorghum genes. These duplications complicate characterizations of microcollinearity and could also interfere with some map-based approaches to gene isolation.

Key words: Genome organization, Collinearity, Tandem duplication, Maize, Sorghum, Rice, Chromosome walking

INTRODUCTION

Mapping studies in sorghum using DNA markers derived from maize indicated that these two grasses have a very similar gene content and extensive regions of conserved map order (Hulbert et al., 1990). Subsequent studies extended these comparison across a broad range of cereals, including barley, foxtail millet, rice, rye, sugarcane and wheat (Ahn et al., 1993; Devos et al., 1993; D'Hont et al., 1994; Moore et al., 1995). These data suggested that different grass genomes would be most effectively conceptualized as separate manifestations of a single genome, thus providing the potential for exceptional synergy in the discovery, cloning, study and engineering of important plant genes (Bennetzen and Freeling, 1993).

Implicit in any unified grass genome model was the belief that the common gene content and map collinearity demonstrated at the level of whole genome mapping would also be observed in highly localized comparisons. We have undertaken such studies and found, as expected, that the low-copy-number sequences (presumably genes) along homoeologous 100 kb to 500 kb segments of the maize, rice and sorghum genomes are highly conserved in sequence, position and orientation. The interspersed repetitive DNAs found between these genes (Bennetzen et al., 1994; Springer et al., 1994) are very different, however, making the physical distance between comparable loci highly variable. We have seen a number of tandem and 'distantly tandem' repetitions of several of these genes, which can complicate the use of a parallel genome approach to gene identification, study and isolation.

MICROCOLLINEARITY OF GRASS GENOMES

The *sh2-a1* homoeologous regions

The *sh2* and *a1* loci of maize are tightly linked (0.1 to 0.2 cM) and have been cloned together on a yeast artificial chromosome (YAC) insert. These two genes were found to be about 140 kb apart and transcribed in the same direction, with *sh2* upstream of *a1* (Civardi et al., 1994). We have used the maize *sh2* locus as a probe to isolate bacterial artificial chromosome (BAC) clones containing the genomic region surrounding this locus from rice and sorghum (M. Chen et al., unpublished).

For both rice and sorghum, clones containing *sh2* homologues also contained *a1* homologues. These putative genes were found to be in the same orientation relative to each other in rice and sorghum as they are in maize. However, the distance between these homologues was found to be about 20 kb in both rice and sorghum (M. Chen et al., unpublished).

The *adh1* homoeologous region

We have extensively characterized the DNA around the *adh1* locus of maize (Bennetzen et al., 1994; Springer et al., 1994; Avramova et al., 1995). Although most of the DNA flanking maize *adh1* is repetitive, we have identified a few low-copy-number sequences within 200 kb downstream of *adh1*. Some of these low-copy-number sequences exhibit coding and/or transcriptional properties suggesting that they are genes (Avramova et al., 1995, and unpublished observation). We have found that these low-copy-number, gene-like sequences are also found in the same order in sorghum relative to the sorghum *adh1* homologue. As observed in the *sh2-a1* region, these gene-like sequences are fewer kb apart in sorghum than they are in maize.

SEGMENTAL DUPLICATIONS IN THE MAIZE, RICE AND SORGHUM GENOMES

Duplications of large segments or whole genomes (polyploidy)

Molecular mapping studies have demonstrated that the diploid maize genome probably had a tetraploid origin (Helentjaris et al., 1988; Moore et al., 1995). Similarly, sorghum has been found to have collinear duplications for most of its genomic constitution (Hulbert et al., 1990; Chittenden et al., 1994). Hence, these two species might be best viewed as 'degenerate' tetraploids from the perspective of genome organization.

Additional duplications are also seen in maize and sorghum, and all other plant species, as evidenced by various gene families whose member number shows extensive interspecies variation. In a few instances, tandem gene families in plants have been seen to expand and contract by a process of unequal recombinational exchange (Hulbert and Bennetzen, 1991; Sudupak et al., 1993; Hong et al., 1993). However, in most cases, the origins of these duplications are not known. In rice, one duplication appears to contain nearly an entire chromosome arm (Moore et al., 1995). In other cases, only a few genes from a collinear set appear to be duplicated, often a few cM away on the same chromosome arm. These 'distantly tandem' duplications (Sanz-Alferez et al., 1995) may be the legacies of tandem duplications of large segments or may indicate a *cis* preference for gene movements, like that seen for many transposable elements (which could be the vectors for such gene movement). Regardless of their origin, these duplications can severely interfere with map integration and map-based gene isolation, to name a few possible problems (discussed in more detail by Bennetzen, 1995).

Duplications of *sh2-a1* homologues in sorghum and rice

In our characterizations of a single BAC containing *sh2* and *a1* homologues from sorghum, we found a second *a1* homologue about 10 kb downstream of the first *a1* homologue. These two *a1* homologues of sorghum hybridized with roughly equal intensity to a probe from the maize *a1* locus.

In both *indica* and *japonica* rice, we have identified two separate segments containing linked *sh2* and *a1* homologues. The relative map positions of these two segments has not been determined.

Duplication of *adh1* homologues of sorghum

By both isozymic and DNA sequence criteria, we have determined that sorghum has two homologues of the maize *adh1* locus. These were cloned on separate cosmids, and the two genes are more than 98% homologous in DNA sequence (C.-N. Liu and J. L. Bennetzen, unpublished). A low-copy-number sequence about 15 kb upstream of each of these *adh1* homologues is also about 98% conserved. We have no direct evidence that this duplication is tandem, but mapping studies place the two sorghum *adh1* homologues at approximately the same (homoeologous) location as the single *adh1* locus of maize (Whitkus et al., 1992). The size of this presumed tandem duplication is not known, but our BAC clone data suggest that the two *adh1* homologues are at least 30 kb apart, perhaps much more.

CONCLUSIONS AND OUTLOOK

Both recombinational and physical mapping studies indicate high levels of collinearity and microcollinearity between segments of various grass genomes (Dunford et al., 1995; Killian et al., 1995; M. Chen et al., unpublished). Hence, the assumption that map position can be used to identify and compare 'cross-species' alleles appears to be valid (Bennetzen and Freeling, 1993; Bennetzen, 1995; Paterson et al., 1995). However, tandem and distantly tandem duplications will confuse such comparisons and lead to inappropriate assignments of homoeology (Bennetzen, 1995).

For map-based gene isolation, segmental duplications will be particularly problematic. Steps in an attempted chromosome walk could end up leaping from one segment to another, perhaps unlinked, segment. Leaps to another linkage group could be distinguished by placing each generation of probes on a recombinational map, but this will be very time-consuming (or not definitive) for tandem duplications. Most of these duplications are rare events, so that mutational drift allows the correct step to be discerned from the incorrect step merely by hybridizational intensity. In some cases, however, rapid changes in tandem copy number may be driven by unequal exchange (Sudupak et al., 1993) or drift may be counteracted by gene conversion. In these circumstances, the correct step may be very difficult to differentiate from an incorrect step. As a general rule, perhaps the best approach would be to identify a parallel genome that is not duplicated in an investigator's segment-of-interest before commencing a chromosome walk.

The preparation of this manuscript was supported by a grant from the USDA (94-37300-0299). A. Oliveira and P. SanMiguel are supported by the Brazilian Council of Research and Development (CNPq) and an NSF training grant in Plant Genetics, respectively.

REFERENCES

Ahn, S., Anderson, J. A., Sorrells, M. E. and Tanksley, S. D. (1993). Homoeologous relationships of rice, wheat and maize chromosomes. *Mol. Gen. Genet.* **241**, 483-490.

Avramova, Z., SanMiguel, P., Georgieva, E. and Bennetzen, J. L. (1995). Matrix attachment regions and transcribed sequences within a long chromosomal continuum containing maize *adh1*. *Plant Cell* **7**, 1667-1680.

Bennetzen, J. L. and Freeling, M. (1993). Grasses as a single genetic system: genome composition, collinearity and compatibility. *Trends Genet.* **9**, 259-261.

Bennetzen, J. L., Schrick, K., Springer, P. S., Brown, W. E. and SanMiguel, P. (1994). Active maize genes are unmodified and flanked by diverse classes of modified, highly repetitive DNA. *Genome* **37**, 565-576.

Bennetzen, J. L. (1995). The use of comparative genome mapping in the identification, cloning and manipulation of important plant genes. In *The*

Impact of Plant Molecular Genetics (ed B. W. S. Sobral), pp. 71-85. Birkhauser, Boston.

Chittenden, L. M., Schertz, K. F., Lin, Y.-R., Wing, R. A. and Paterson, A. H. (1994). A detailed RFLP map of *Sorghum bicolor* × *S. propinquum*, suitable for high-density mapping, suggests ancestral duplication of *Sorghum* chromosomes or chromosomal segments. *Theor. Appl. Genet.* **87**, 925-933.

Civardi, L., Xia, Y., Edwards, K. J., Schnable, P. S. and Nikolau, B. J. (1994). The relationship between genetic and physical distances in the cloned *a1-sh2* interval of the *Zea mays* L. genome. *Proc. Nat. Acad. Sci. USA* **91**, 8268-8272.

Devos, K. M., Millan, T. and Gale, M. D. (1993). Comparative RFLP maps of homoeologous group 2 chromosomes of wheat, rye, and barley. *Theor. Appl. Genet.* **85**, 784-792.

D'Hont, A., Lu, Y.-H., de Leon, D. G., Grivet, L., Feldmann, P., Lanaud, C. and Glaszmann, J. C. (1994). A molecular approach to unraveling the genetics of sugarcane, a complex polyploid of the Andropogoneae tribe. *Genome* **37**, 222-230.

Dunford, R. P., Kurata, N., Laurie, D. A., Money, T. A., Minobe, Y. and Moore, G. (1995). Conservation of fine-scale DNA marker order in the genomes of rice and the Triticeae. *Nucl. Acids Res.* **23**, 2724-2728.

Helentjaris, T., Weber, D. L. and Wright, S. (1988). Identification of the genomic locations of duplicate nucleotide sequences in maize by analysis of restriction fragment length polymorphisms. *Genetics* **118**, 353-363.

Hong, K. S., Richter, T. E., Bennetzen, J. L. and Hulbert, S. H. (1993). Complex line-specific duplications in maize. *Mol. Gen. Genet.* **239**, 115-121.

Hulbert, S. H., Richter, T. E., Axtell, J. D. and Bennetzen, J. L. (1990). Genetic mapping and characterization of sorghum and related crops by means of maize DNA probes. *Proc. Nat. Acad. Sci. USA* **87**, 4251-4255.

Hulbert, S. H. and Bennetzen, J. L. (1991). Recombination at the *Rp1* locus of maize. *Mol. Gen. Genet.* **226**, 377-382.

Kilian, A., Kudrna, D. A., Kleinhofs, A., Yano, M., Kurata, N., Steffenson, B. and Sasaki, T. (1995). Rice-barley synteny and its application to saturation mapping of the barley Rpg1 region. *Nucl. Acids Res.* **23**, 2729-2733.

Moore, G., Devos, K. M., Wang, Z. and Gale, M. D. (1995). Grasses, line up and form a circle. *Curr. Biol.* **5**, 737-739.

Paterson, A. H., Lin, Y.-R., Li, Z., Schertz, K. F., Doebley, J. F., Pinson, S. R. M., Liu, S.-C., Stansel, J. W. and Irvine, J. E. (1995). Convergent evolution of cereal crops by independent mutations at corresponding genetic loci. *Science* **269**, 1714-1717.

Sanz-Alferez, S., Richter, T. E., Hulbert, S. H. and Bennetzen, J. L. (1995). The *Rp3* disease resistance gene of maize: mapping and characterization of introgressed alleles. *Theor. Appl. Genet.* **91**, 25-32.

Springer, P. S., Edwards, K. J. and Bennetzen, J. L. (1994). DNA class organization on maize *Adh1* yeast artificial chromosomes. *Proc. Nat. Acad. Sci. USA* **91**, 863-867.

Sudupak, M. A., Bennetzen, J. L. and Hulbert, S. H. (1993). Unequal exchange and meiotic instability of *Rp1* region disease resistance genes in maize. *Genetics* **133**, 119-125.

Whitkus, R., Doebley, J. and Lee, M. (1992). Comparative genome mapping of sorghum and maize. *Genetics* **132**, 1119-1130.

Sequencing and mapping the *Arabidopsis* genome: a weed model for real crops

Michel Delseny, Monique Raynal, Michele Laudié, Fabrice Varoquaux, Pascale Comella, Hui Ju Wu, Richard Cooke and Francoise Grellet

Physiologie et Biologie Moléculaire des Plantes, CNRS Unité 565, GDR 1003, University of Perpignan, 66860 Perpignan, France

SUMMARY

Arabidopsis is a crucifer weed with a small genome of about 120 Mbp which has been chosen as a model species for plant molecular genetics. Four years ago, a consortium of nine French laboratories, including ours, initiated a project aimed at mapping the transcribed regions of the genome. The strategy employed was to systematically and randomly sequence cDNA clones isolated from libraries made from different tissues and organs of plants grown under various physiological conditions. The consortium released about 7,000 expressed sequenced tags (ESTs) in the dbEST database corresponding to approximately 3,500 unique genes. In the next phase of the programme, a YAC library with average inserts of 500 kbp has been prepared. We have now started to use the EST information to map the cDNA clones on these YACs. The most recent aspect of *Arabidopsis* sequencing is the ESSA (European Scientists Sequencing *Arabidopsis*) project, in which the aim is to describe 2.5 Mbp by the end of 1996. Genomic sequencing has revealed a very high gene density. Comparison of present genomic sequencing results with the EST data suggests that up to half of the genes might already be tagged with an EST. In collaboration with Carlos Quiros' group in Davis we have also analysed the conservation of a 30 kbp locus (*Em* 1, a late embryogenesis abundant protein gene) on chromosome 3 between *Arabidopsis* and several *Brassica* species. Progress on these various aspects will be reviewed. We shall also present some sequence comparisons between *Arabidopsis* and rice ESTs. These results suggest that it should be possible in the very near future to map a pool of common genes onto many different plant genomes. This should provide a common framework to integrate maps from different species and facilitate map-based cloning of genes of agronomical importance.

Key words: *Arabidopsis*, EST, Genome sequencing, Mapping

INTRODUCTION

The genome size of cultivated plant species varies from a few hundred megabases for rice or cabbage to more than ten thousand megabases for wheat or oat. However, the gene number is likely to be approximately the same in all plant species, in the range of 20-50,000. Many of these genes have identical or similar functions and, despite the degeneracy of the genetic code, one expects that a large proportion of the genes will be conserved between different plant species. With these considerations in mind, many groups around the world decided to use a model plant to decipher the organisation and function-ing of the plant genome. The *Arabidopsis thaliana* weed species was chosen because of its small genome, now estimated to be 120 Mbp, its short life cycle, the relative ease of mutagenesis and genetic analysis and because it is amenable to genetic trans-formation and reverse genetics. For reviews on *Arabidopsis* and its genome see Koncz et al., 1992; Meyerowitz and Somerville, 1994; Delseny and Cooke, 1996. Despite this small genome, isolating and identifying all the genes from this species using classical strategies remains a daunting task. Therefore, four years ago, with a group of colleagues, we set up a consortium supported initially by CNRS, the main French Research Council. Nine groups, belonging to CNRS, INRA or Universi-ties, and located in Toulouse, Versailles, Orsay, Gif, Stras-bourg, Grenoble and Perpignan were associated in a 'groupe de recherches' (GDR 1003) with the aim of systematically sequencing extremities of randomly selected cDNA clones, identifying as many of them as possible by homology with sequences from other organisms already deposited in databases, and finally mapping these genes onto the genetic map. Since the end of 1993 the sequencing programme has been mainly supported by the European Community as part of the ESSA project with the requirement that only new sequences will be paid for and with the goal to produce 3,000 new ESTs. This project has been outstandingly successful and this report will review some of its achievements and applications. The latter are in two major fields: sequencing large portions of *Arabidopsis* chromosomes and transferring the currently available informa-tion to the analysis of the plant genome and the characterisa-tion of agronomically important genes from cultivated crops.

SEQUENCING cDNA

Orientated libraries have been prepared from a variety of tissues and physiological situations, including cell suspension cultures, wounded leaves, elicitor treated cells, young green in vitro cultured plants, etiolated seedlings, flower buds, immature siliques and dry seeds. From each library several hundred cDNAs were randomly selected and sequenced. Sequences were routinely aligned against each other, thanks to a central database in Toulouse, and, when the concurrent American programme started to release data, against the American EST collection. Our strategy differs from the one adopted in the USA in the sense that we are using multiple libraries, and that whenever a clone is considered to be new, we usually sequence the other extremity. As a result of this effort we have now carried out more than 10,000 single sequencing runs and released about 7,000 ESTs in EMBL and dbEST databases corresponding to nearly 5,000 independent clones, of which 2,100 have been sequenced from both ends. Since the beginning of ESSA, 1,800 ESTs corresponding to 900 new clones have been validated for funding. Approximately 5,000 of our released ESTs correspond to non-redundant sequences, thus tagging about 3,000 different genes. Together with the work at Michigan State University, more than 20,000 EST are now freely available and this probably represents between one third and one half of the *Arabidopsis* expressed genes. As shown in Table 1, this has placed *Arabidopsis* in second position, far behind the human genome project but ahead of the *Caenorhabditis elegans* and rice genomes. However, after the rapid initial progress in the French and American programmes, it is becoming increasingly more difficult and expensive to produce novel EST sequences. Consequently several groups have already started to eliminate the most abundant clones from their libraries and it can be anticipated that this cDNA systematic and partial sequencing programme will terminate by the end of ESSA next year. Analyses of these sequences were reported both by the French groups (Hofte et al., 1993; Cooke et al., 1996) and by the American group (Newman et al., 1994). By using classical alignment programmes, we could putatively identify about one third of the non-redundant sequences with previously known genes and proteins in other species.

This result indicates that either many genes are specific to plants or that they code for proteins which have similar functions in other organisms but which have diverged to such an extent that they cannot be recognised as similar by our alignment algorithms. A bias in this type of analysis is that we can compare only sequences at the ends of molecules whereas the conserved regions are frequently in the core of the proteins, close to catalytic sites. Another point of interest is that several of our unidentified ESTs match unidentified ESTs in other organisms. This category of EST corresponds to relatively abundantly expressed and conserved genes but which have not yet been assigned a function.

APPLICATIONS TO BIOLOGICAL QUESTIONS

Many of these sequences have immediately been used either by ourselves or by colleagues in order to analyse biological problems, for example the elucidation of metabolic or signal transduction pathways. Several such examples have already been reported, including that of the regulation of expression of late embryogenesis abundant protein genes (Parcy et al., 1994), the identification and expression of E2-type ubiquitin conjugating enzymes (Genschik et al., 1994), the characterisation of thioredoxin h and thioredoxin reductases (Jacquot et al., 1994; Rivera-Madrid et al., 1995) or the characterisation of several GA-20 oxidase genes (Phillips et al., 1995). Another important point is that the homology which is observed between ESTs and other genes is just an indication but not a proof of the function of the gene product. In most of the cases, the rigourous identification of the function remains to be determined by in vivo or in vitro experimentation. One of the most significant results of this study was the identification of an unexpected large number of multigene families. This discovery of a high level of genetic redundancy in such a small genome implies a limitation of the efficiency of gene disruption strategies to identify and characterise gene products.

Two other interesting aspects have been developed essentially by colleagues. One is the search for microsatellites which can be used for mapping purposes. The results of screening cDNA clones isolated from a silique library with three different oligonucleotide probes, followed by partial sequencing of hybridising clones revealed a number of microsatellite sequences in transcribed regions (Table 2); preliminary results were recently published (Depeiges et al., 1995). Another example of a search for simple sequences was reported by Regad et al. (1994); these authors have examined 2,660 EST and found that the telomere sequence AAACCCTAA is overrepresented in transcribed sequences. They proposed a model for explaining the propagation of this sequence which suggests that a telomerase may be involved in repairing ends of DNA strands following a recombination event. A consequence is that these dispersed telomere motifs, which have also been detected in other species, might be useful as general markers and may represent footprints of recombination events.

MAPPING THE EST

From the very beginning of the programme the mapping of expressed genes was a major objective. This seemed to us

Table 1. dbEST release (08.17.95)

Homo sapiens	233,151
Arabidopsis thaliana	20162
Caenorhabditis elegans	12,102
Oryza sativa	10,990
Saccharomyces cerevisiae	2,944
Zea mays	1,173
Ricinus communis	750
Brassica campestris	645
Brassica napus	350
Solanum tuberosum	49

Table 2. Examples of microsatellites in cDNA

(ATG)n		
G (TGA)7	3′ NC	Calmodulin 1
TGT (TGA)6TCTTGAGGA	3′ NC	Cyclophilin
(AAG)n		
AC(CTT)5 CG	C	3 Ketoacyl- CoA -thiolase
T(TCT)5CCTT	3′ NC	?
CAT(CTT)5CGT	5′ NC	S13 ribosomal protein
AAA(AAG)5	5′ NC	?
(CAT)n		
(CAT)6CAG	3′ NC	ATP-binding protein
(AG)n		
(TC)2CC(TC)9GTTTTCG(TC)4	3′NC	?

important because it should considerably facilitate positional cloning of genes which are just identified by a mutation and also help in YAC contig assembly. This aspect of the programme has been delayed because sequencing was a priority and because it was decided to prepare a new YAC library. This library known as the CIC (CNRS, INRA, CEPH) library is the result of the collaboration of several groups in the GDR 1003 consortium and Daniel Cohen's group at the CEPH (Centre d'Etude du Polymorphisme Humain). It is characterised by large (~500 kbp) and non chimeric insertions (Creuzot et al., 1995) and it has been released to a limited number of groups recently in order to complete the physical map. As a result several large contigs have been ordered and the map of chromosome 4 is soon to be published (Schmidt et al., 1995). The strategy for mapping cDNA was initially restriction fragment length polymorphism (RFLP) based, using recombinant inbred lines (Lister and Dean, 1993). However, this approach was inefficient since we found polymorphisms with only 30% of the clones and it was inappropriate for multigene families. Nevertheless, about 50 clones have been mapped using this technique. With the availability of the ordered CIC YAC library, an alternative strategy was feasible. DNA was prepared from different combinations of YAC pools and used in PCR reactions with specific primers designed from the ESTs; in this way it is possible to determine from which clone a fragment of the expected size has been amplified and to assign to this clone the corresponding cDNA. Having sequence from both ends of a cDNA is a considerable advantage in the detection of polymorphisms and identification of individual members of a multigene family. In this process, the amplification product is compared to a control using genomic DNA and the same primers. These control experiments are carried out with the DNA extracted from the parental lines of the recombinant inbred lines. Following amplification we systematically digest the product with a dozen 5 and 4 bp cutter restriction enzymes in order to detect a polymorphism. Whenever a polymorphism is observed, DNA from each RI line is amplified and digested with the appropriate enzyme. As a result the YAC clones identified by PCR can be anchored on the genetic map. The flow-chart of the mapping steps is presented in Table 3. During the last 6 months about 120 cDNAs have thus been positioned on either a YAC or on the genetic map or both and this number is expected to increase rapidly in the next few months. As a first step, the groups participating in this project are mapping as a priority the identified genes for which they have a scientific interest rather than randomly selected clones. In our group, we are primarily mapping genes expressed during late embryogenesis as well as genes coding for proteins of the translation and transcription machineries. This mapping strategy has several advantages: it facilitates the detection of chimeric YAC clones and the assembly of YAC contigs. As a consequence this work should accelerate physical mapping of the genome and result in the addition of many expressed sequences as landmarks on the map, which is more satisfactory than anonymous sequences.

GENOMIC SEQUENCING

Our laboratory is also participating in the ESSA project for genomic sequencing. The goal of the overall project is to

Table 3. Strategy for mapping EST

(1) PCR amplify genomic DNA from Columbia and Landsberg ecotypes

(2) Search for polymorphism using 12 four base cutter restriction enzymes: CAPS

(3) If polymorphism, screen 100 RI Lines and map, then PCR amplify YAC pools

(4) If no CAPS polymorphism, digest parental DNA with 12 six base cutter restriction enzymes, search RFLP, and map

(5) PCR amplify YAC pools and identify corresponding clones

(6) Examine whether the YAC is already anchored or not. If not search for other polymorphic sequences on the same YAC

(7) Organize contigs

analyse about 2.5 Mbp of genomic DNA located in two large contigs on chromosome 4 (2 Mbp) and in several additional loci around genes of interest (0.5 Mbp). We are contributing to the sequencing of a 75 kbp region centered on the late embryogenesis abundant protein gene *Em 1* (Gaubier et al., 1993) which is located on chromosome 3, close to the *APETALA 3* gene. We have already determined about 30 kbp and have identified 7 putative genes in addition to *Em1* which suggest that, in this region, the gene density is very high with a gene approximately every 4 kbp. Genes were identified using several approaches. Initially short segments were aligned against data bases using FASTA or BLAST algorithms in order to find discrete homologies; then the homology was extended progressively to the adjacent sequences. The use of existing programmes such as GRAIL, or GENEFINDER was totally unadapted to the detection of genes in *Arabidopsis*. When we were reasonably confident that a gene was present, we used the longest open reading frame in the region as a probe to screen various cDNA libraries and identify the cognate cDNA. Full length cDNA clones were isolated for three genes flanking *Em*. This allowed us to precisely define the intron and exon sequences. With this information as well as with similar data originating from B. Lescure's lab in Toulouse, it becomes possible for S. Klostermann at MIPS in Martinsried to modify the existing search programmes in order to enhance the probability of detection of genes in *Arabidopsis*. As both the genomic sequencing and the EST programmes progress, it becomes possible to use the improved version of GENEFINDER as well as the homologous EST sequences from *Arabidopsis* and from rice (Sasaki et al., 1994) to determine the limits of the genes. Our results of genomic sequencing are summarised in Table 4 and characterization of a gene immediately adjacent to *Em 1* has recently been reported (Gaubier et al., 1995). None of these genes have been previously described in plants. Only one does not present any homology with a known gene; however, part of its sequence shows 89% homology at the protein level with an unidentified maize EST. Three of the genes perfectly match the *Arabidopsis* EST sequences and to a lesser extent the EST sequences from other species. No homologous plant EST was found for the three remaining genes but clear similarities in conserved motifs with animal or bacterial sequences gave a hint of their function. This type of information is essentially confirmed by results of other ESSA sequencers. Close to 500 kbp on chro-

Table 4. Genes identified in the *Em* locus (23,080 bp)

Gene 1	Unknown gene, 250 amino acids, probably 3 introns; homology with unknown maize EST (89%), more distantly related to rice and *Arabidopsis* ESTs
Gene 2	Unknown gene, 256 amino acids including signal peptide, no intron Cognate cDNAs Homology with *Escherichia coli* ORF (61%) and *Bradyrhizobium* gene *Cyc J* (58%)(cytochrome C synthesis?) No EST yet
Gene 3	Unknown gene, 392 amino acids, 8 introns Cognate cDNA Homology with rat p67 protein: inhibitor of translation initiation factor phosphorylation (59%), with yeast p67 (62%) *Arabidopsis* EST (100%); however, other *Arabidopsis* ESTs have higher homology with rat p67, human, mouse, *C. elegans* and rice ESTs
Gene 4	*Em 1* gene , 152 amino acids, 1 intron Cognate cDNAs Homologues in maize, barley, cotton, soybean...
Gene 5	Unknown gene, 387 amino acids, 13 introns Cognate cDNA Homology with *Rhodobacter capsulatus* Bch G gene: chlorophyll synthase (61%) *Arabidopsis* EST (100%) (Gaubier et al., 1995)
Gene 6	Unknown gene, 507 amino acids, 17 introns Homology with yeast *Sac I* gene (Sec 14 suppressor) *Arabidopsis* EST (100%), also rice and *C. elegans* ESTs
Gene 7	Unknown gene , uncomplete Homologies with Acyl CoA deshydrogenases No EST yet

mosome 4 and 230 kbp in other regions have now been sequenced and analysed: about half of the predicted genes match with ESTs from *Arabidopsis* or other species and on average there is one gene every 4-5 kbp. Assuming a genome size of 120 Mbp, and a homogenous gene density similar to the one observed, allowing for 20% of repeated sequences including rDNA, satellite DNA, telomeres and a few transposable elements, there is sufficient space for no more than 19-24,000 protein coding genes in the *Arabidopsis* genome.

STRATEGIES FOR TRANSFERRING *ARABIDOPSIS* GENOME INFORMATION TO OTHER CROPS

There are several ways to use the *Arabidopsis* genome information to analyse other crop genomes, depending on the final goal.

Use of *Arabidopsis* cDNA sequence as probes

A first striking point is that there is enough conservation at the amino acid sequence level for a large number of genes to be recognized simply on the basis of homology. In our analysis of the cDNA sequences we could identify a large number of the so-called ancient conserved regions, or ACR (Claverie, 1993), which correspond to highly conserved motifs in all living organisms. However, ACR are difficult to use for mapping purposes and we searched for another approach. The most obvious strategy for comparative mapping is to use a set of common probes in RFLP experiments. To do that one needs highly conserved sequences. In order to define such a set of probes we aligned approximately 4,000 non-redundant *Arabidopsis* ESTs with the available rice ESTs in dbEST using a

BLAST N score >400 which is equivalent to homology at the nucleotide level higher than 90% over 200 bp. We obtained a series of about 300 clones coding for highly conserved genes which will probably cross-hybridise with most other dicot species and almost certainly with a large number of monocots. As more sequences are screened, additional candidates should be found. This set of probes should be used firstly as a set of common landmarks on all genomes and as such they should be very helpful in analysing the conservation and evolution of plant genomes. An alternative approach is to try to develop universal primers for all species and to use a CAPS strategy. Interestingly, most of these ESTs correspond to putatively identified sequences and several of them are already mapped on the rice genome (Kurata et al., 1994). Examples of such genes are presented in Table 5. In addition to these universal markers, many ESTs identify similar proteins in rice and *Arabidopsis* but the sequence has too much divergence to be used directly in hybridisation experiments; however, some domains are conserved and can be used to design PCR primers which should allow the amplification of the corresponding DNA sequence in many different species. The amplified fragment can be used directly as an RFLP probe or it can be digested with restriction enzymes until a polymorphism is found. Again such a strategy should allow one to map homologous genes in various species.

Finally, for many genes, the homology is too low, even in conserved protein domains, to engineer specific primers useful for a variety of different species. However, the conservation should be sufficiently high to allow hybridisation to or amplification with the DNA of a closely related species. For example about 90% of *Arabidopsis* genes recognise an homologue in *Brassica* (Lagercrantz et al., 1996).

Search for synteny between *Arabidopsis* and other crops

Use of a set of common probes should facilitate search for synteny between the *Arabidopsis* genome and other crops. Even if there is evidence for synteny (Kowalski et al., 1994; Lagercrantz et al., 1996) it is difficult to evaluate it on a large

Table 5. Examples of genes with highly conserved nucleotide sequences between *Arabidopsis* and rice

Adenylate translocator	GA 3 PDH
Ubiquitin	Actin
HSP70 cognate	EF1 alpha
SAM synthase	Beta-tubulin
Actin depolymerizing factor	Poly(A) binding protein
Histone H4	Calmodulin 1
Cystein protease inhibitor	Thiol protease
Vacuolar H+ pyrophosphatase	LHCP
Glutamine synthase	EIF 4A2
EF 2	ADH I
Ubiquitin conjugating enzyme E2	Alpha-tubulin
Enolase	Sucrose synthase
Proteasome alpha subunit	Triose phosphate isomerase
Mito fusion target protein	Adenosyl homocystein synthase
Histone H2B	Tonoplast Intrinsic protein
Ribosomal protein S4	S12
S13	S14
S16	Ribosomal protein L1A
L2	L3
L5	L18
L19	L21
L30 A	L32
L46	

scale when a limited number of probes are being used. Recently, in collaboration with Carlos Quiros' group in Davis, we took advantage of our knowledge at the sequence level of the organisation of the *Em 1* locus. We provided genomic DNA fragments, as well as cDNA probes, to our colleagues and they examined their organization and arrangements in the three diploid *Brasssica*: *B. campestris*, *B. oleracea* and *B. nigra*. For each of the *Brassica* genomes they found that the cluster of 5 genes is conserved on a single linkage group and apparently in the same order. However, partial complexes were found duplicated in other chromosome segments on the same or different linkage groups. These data establish that the *Brassica* chromosomes have undergone extensive duplications and rearrangements during the course of their evolution. A similar situation is expected in many dicot crop species and this might obscure synteny between genomes. However, a local synteny almost certainly exists and the size of such synteny blocks will have to be determined. The availability of a physical map of chromosome 4 which will contain many identified markers should be an excellent tool to determine how the sequences in this chromosome are distributed in various related or more distant species. If the existence of local synteny is proven then it may be appropriate to use *Arabidopsis* as an intermediate cloning step to isolate genes of interest in other species.

CONCLUSION

The *Arabidopsis* genome is now clearly the most extensively analysed plant genome. Due to intensive effort and collaboration more than 20,000 ESTs and many corresponding clones are freely available, a physical map is already available for one chromosome and maps for each of the four other chromosomes should be available in the foreseeable future. Finally it is likely that the first plant chromosome to be sequenced will be an *Arabidopsis* one. Although a lot of experiments remains to be carried out in order to complete the physical and genetical map and despite the fact that close to two thirds of the genes are awaiting to be assigned a function, tremendous progress has been made in just the last few years. As a result of this effort it is becoming possible to design strategies to use the *Arabidopsis* genome information in order to analyse the genome of the economically important crops. Preliminary results indicate that we have now in hand most of the tools necessary to attempt a unifying of plant genomes and thus to understand how the genome of each species has evolved from a common ancestor.

This work was supported by CNRS, GREG and EEC grants to our group as well as to participants of the various networks. The authors thank all their colleagues from these French or European networks, as well as C. Quiros and J. Sadowski in Davis, for providing unpublished information and for stimulating discussions.

REFERENCES

Claverie, J. M. (1993). Database of ancient sequences. *Nature* **364**, 19-20.

Cooke, R., Raynal, M., Laudié, M., Grellet, F., Delseny, M., Morris, P. C., Guerrier, D., Giraudat, J., Quigley, F., Clabault, G., et al. (1996). Further progress towards a catalogue of all *Arabidopsis* genes: analysis of a set of 5000 non redundant ESTs. *Plant J.* **9**, 101-124.

Creusot, F., Fouilloux, E., Dron, M., Lafleuriel, J., Picard, G., Billault, A., Le Paslier, D., Cohen, D., Chabouté, M. E. and Durr, A. (1995). The CIC library: a large insert YAC library for genome mapping in *Arabidopsis thaliana*. *Plant J.* **8**, 763-770.

Delseny, M. and Cooke, R. (1996). The *Arabidopsis* nuclear genome. In *Handbook of Genome Analysis* (ed. W. Bodmer, N. K. Spurr, N. Teich, C. Rawlings and B. D. Young) Blackwell Scientific (in press).

Depeiges, A., Goubely, C., Lenoir, A., Cocherel, S., Picard, G., Raynal, M., Grellet, F. and Delseny, M. (1995). Identification of the most represented motifs in *Arabidopsis thaliana* microsatellite loci. *Theor. Appl. Genet.* **91**, 160-168.

Gaubier, P., Raynal, M., Hull, G., Huestis, G. M., Grellet, F., Arenas, C., Pages, M. and Delseny, M. (1993). Two different Em-like genes are expressed in *Arabidopsis thaliana* seeds during maturation. *Mol. Gen. Genet.* **238**, 409-418.

Gaubier, P., Wu, H. J., Laudié, M., Delseny, M. and Grellet, F. (1995). A chlorophyll synthetase gene from *Arabidopsis thaliana*. *Mol. Gen. Genet.* **249**, 58-64.

Genschik, P., Durr, A. and Fleck, J. (1994). Differential expression of several E2-type ubiquitin carrier protein genes at different development stages in *Arabidopsis thaliana* and *Nicotiana sylvestris*. *Mol. Gen. Genet.* **244**, 548-556.

Hofte, H., Desprez, T., Amselem, J., Chiapello, H., Caboche, M., Moisan, A., Jourjon, M. F., Charpenteau, J. L., Berthomieu, P., Guerrier, F., et al. (1993). An inventory of 1152 expressed sequence tags obtained by partial sequencing of cDNA from *Arabidopsis thaliana*. *Plant J.* **4**, 1051-1061.

Jacquot, J. P., Rivera-Madrid, R., Marinho, P., Kollarova, M., Le Marechal, P., Miginiac-Maslow, M. and Meyer, Y. (1994). *Arabidopsis thaliana* NAPDH – thioredoxin reductase cDNA. Characterization and expression of the recombinant protein in *E. coli*. *J. Mol. Biol.* **235**, 1357-1363.

Koncz, C., Chua, N. H. and Schell, J. (1992). *Methods in* Arabidopsis *Research*. World Scientific, Singapore.

Kowalski, S. P., Lan, T. H., Feldman, K. A. and Paterson, A. H. (1994). Comparative mapping of *Arabidopsis thaliana* and *Brassica oleracea* chromosomes reveals islands of conserved organization. *Genetics* **138**, 499-510.

Kurata, N., Nagamura, Y., Yamamoto, K., Harushima, Y. H., Sue, N., Wu, J., Antonio, B. A., Shomura, A., Shimizu, T., Lin, S. Y., et al. (1994). A 300kb interval rice genetic map including 883 expressed sequences. *Nature Genet.* **8**, 365-372.

Lagercrantz, U., Putterill, J., Coupland, G. and Lydiate, D. (1996). Comparative mapping in Arabidopsis and Brassica, fine scale genome collinearity and congruence of genes controlling flowering time. *Plant J.* **9**, 13-20.

Lister, C. and Dean, C. (1993). Recombinant inbred lines for mapping RFLP and phenotypic markers in *Arabidopsis thaliana*. *Plant. J.* **4**, 745-750.

Meyerowitz, E. and Somerville, C. (1994). Arabidopsis. Cold Spring Harbor Laboratory Press, Cold Spring Harbor, NY.

Newman, T., De Bruijn, F., Green, P., Keegstra, K., Kende, H., Mc Intosh, L., Ohlrogge, J., Raikhel, N., Somerville, S., Thomashow, M., Retzel, E. and Somerville, C. (1994). Genes galore: a summary of the methods for accessing the result of large scale partial sequencing of anonymous *Arabidopsis thaliana* cDNA clones. *Plant Physiol.* **106**, 1241-1255.

Parcy, F., Valon, C., Raynal, M., Gaubier-Comella, P., Delseny, M. and Giraudat, J. (1994). Regulation of gene expression programme during *Arabidopsis* seed development: roles of the *ABI 3* locus and of exogenous abscisic acid. *Plant Cell* **6**, 1567-1582.

Philipps, A. L., Ward, D. A., Uknes, S., Appleford, N. E. S., Langé, T., Huttly, A. K., Garkin, P., Graebe, J. E. and Hedden, P. (1995). Isolation and expression of three Gibberellin 20-oxidase cDNA clones from *Arabidopsis*. *Plant Physiol.* **108**, 1049 1057.

Regad, F., Lebas, M. and Lescure, B. (1994). Interstitial telomeric repeats within the *Arabidopsis thaliana* genome. *J. Mol. Biol.* **239**, 163-169.

Rivera-Madrid, R., Mestres, D., Marinho, P., Jacquot, J. P., Decottignier, P., Miginiac-Maslow, M. and Meyer, Y. (1995). Evidence for five divergent thioredoxin h sequences in *Arabidopsis thaliana*. *Proc. Nat. Acad Sci. USA* **92**, 5620-5624.

Sadowski, J., Gaubier-Comella, P., Delseny, M. and Quiros, C. F. (1996). Genetic and physical mapping of a gene cluster from *Arabidopsis thaliana* in *Brassica* diploid species. *Mol. Gen. Genet.* **251**, 298-306.

Sasaki, T., Song, J. Y., Koga Ban, Y., Matsui, E., Fang, F., Higo, H., Nagasaki, H., Hori, M., Miya, M., Murayama-Kayano, E., et al. (1994). Towards cataloguing all rice genes: large scale sequence of randomly chosen rice cDNA from a callus cDNA library. *Plant J.* **6**, 615-624.

Schmidt, R., West, J., Love, K., Lenehan, Z., Lister, C., Thompson, H., Bouchez, D. and Dean, C. (1995). A physical map of *Arabidopsis thaliana* chromosome 4. *Science* **270**, 480-484.

Printed in Great Britain © The Society for Experimental Biology 1996
SEB0024

Rice cDNAs as a model for expressed genes of plants

Takuji Sasaki

Rice Genome Research Program (RGP), NIAR/STAFF, 1-2, Kannondai 2-chome, Tsukuba, Ibaraki 305, Japan

(e-mail: tsasaki@abr.affrc.go.jp)

SUMMARY

Large-scale rice cDNA analysis has produced a huge amount of nucleotide sequence information for expressed genes in rice. The genes of cDNA clones putatively identified by similarity search were originally found in many different organisms. However, genes identified at a higher confidence level were found in plants, especially in monocots. This means the sequence information produced in random cloning of rice cDNA is useful for the study of other *Gramineae*. Further, assigned gene names of cDNAs mapped on linkage group 6 were grouped by their original species. The functions of gene products for 51% of mapped cDNAs were assigned and 67% of them were known in plants. The map information obtained by linking the position of a cDNA locus and its assigned gene function is indispensable for elucidating collinearity of genes among plant genomes.

Key words: cDNA, Similarity search, RFLP, Rice

INTRODUCTION

The Rice Genome Research Program (RGP) that started in 1991 as a collaborative research program between the National Institute of Agrobiological Resources (NIAR) and the Society of Techno-innovation for Agriculture, Forestry and Fisheries (STAFF) has randomly cloned and partially sequenced more than 25,000 rice cDNAs from several kinds of cDNA libraries, such as green shoot, etiolated shoot, root and callus cultured in the presence of 2,4-dichlorophenoxyacetic acid (2,4-D), benzyladenine or gibberellin. A similarity search of the translated amino acid sequences using the FASTA algorithm against a public database, PIR (Protein Information Resources), revealed that about 25% of them showed significant sequence similarity to amino acid sequences of known proteins, not only from plants, but from a wide variety of organisms (Sasaki et al., 1994). This indicates the possibility that, when these cDNA clones are used as RFLP markers for linkage analysis, the determined loci correspond to the location of the gene for the assigned functional product.

RGP is also engaged in constructing an RFLP linkage map using 186 plants of an F_2 population produced by crossing *japonica* and *indica* varieties (Kurata et al., 1994b). Up to now, about 1,300 cDNAs have been mapped among the 1,904 DNA markers on our genetic map. Our map is used to tag a phenotypical trait accurately, because the average distance between each marker is less than 1 cM. Another promising use of our many DNA markers is as RFLP markers for other *Gramineae* genomes. Cross-checking of the loci obtained by the same markers enables one to clarify the similarity of gene order in the grass family. The genome size of rice is the smallest among major cultivated grasses and, if the similarity of a gene sequence or synteny is established, the allelic gene among grasses could be isolated using the rice physical map.

Physical map construction is under way using YAC (yeast artificial chromosome) as a vector to carry rice genomic DNA fragments with an average size of 350 kb. DNA markers on the linkage map are used as probes for screening the YAC library containing 7,000 clones (Umehara et al., 1995). The conversion of a map based on genetic recombination to a physical alignment means the linking of RFLP markers. If putatively identified cDNA clones other than those already used as RFLP markers are also associated with a specific YAC clone, we then know the location of genes for the assigned functional product. This information should be applicable to other grass species, too.

This report shows, first, how the degree of sequence similarity of rice cDNAs to other organisms is elucidated, and second, how the sequential alignment of putatively identified protein names on a linkage map is made.

GROUPING OF PUTATIVELY IDENTIFIED RICE cDNAs

A partial nucleotide sequence of each cDNA clone from callus with 2,4-D or root library was translated to its amino acid sequence for all three reading frames from the 5′-end. Because of the asymmetric construction of cDNAs with *Sal*I and *Eco*RI adaptors, the direction of the cDNA was unambiguous. Such an amino acid sequence was used as a query sequence to search similar sequences in the PIR database by the FASTA algorithm (Pearson and Lipman, 1988). In cases where the top ranked sequence showed similarity to a query sequence with an optimized score of more than 200, the protein name of this sequence was putatively given to the name of the rice query sequence. From 2,4-D callus and root cDNA libraries we could identify 279 and 291 independent sequences with assigned protein product names, respectively.

12   T. Sasaki

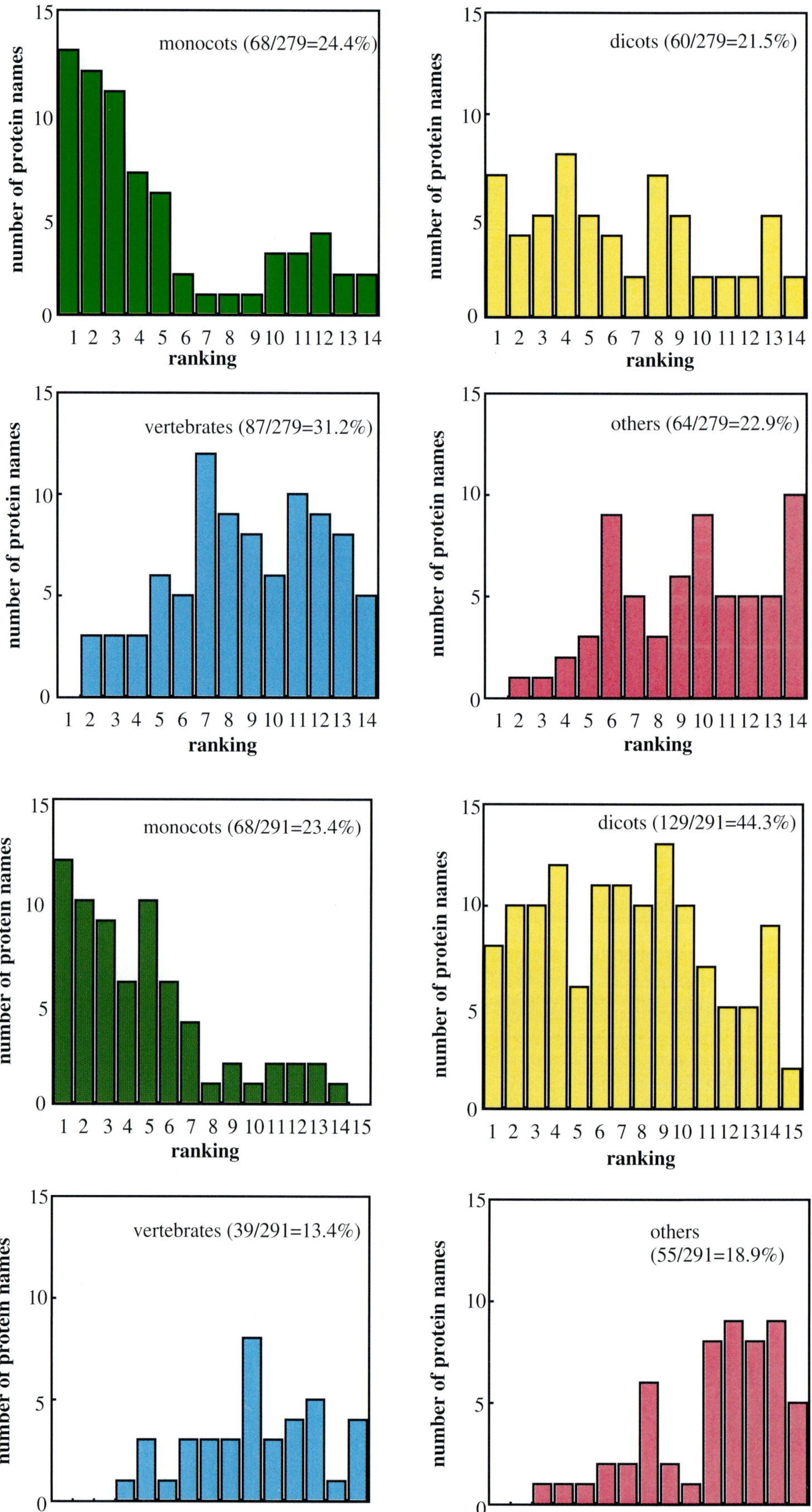

Fig. 1. Categorized cDNA clones with a significant similarity hit in a cDNA library of callus in the presence of 2,4-D. Identified cDNA clones were ranked in order of their degree of matching to each other and twenty protein names from the top of each rank (not shown) were grouped into monocot plants, dicot plants, vertebrates, and others.

Fig. 2. Categorized cDNA clones with a significant similarity hit in a root cDNA library. The grouping procedure is the same as adopted in Fig. 1.

```
protein Z1:102 ANVFVDSSLKLKPSFKDLVVGKYKGETQSVDFQTKAPEVAGQVNSWVEKITTGLIKEILP 161
               + ++    +  + P+    V    Y+G  +SV+FQT A +   G  +N+WVE  T G+I+  IL
ovalbumin: 103 SRLYAQETYTVVPEYLQCVKELYRGGLESVNFQTAADQARGLINAWVESQTNGIIRNILQ 162


protein Z1:162 AGSVDSTTRLVLGNALYFKGSWTEKFDASKTKDEKFHLLDGSSVQTPFMSSTKKQYISSY 221
               SVDS T +VL NA+ FKG W + F  A  T+    F ++   S      M        ++S
ovalbumin: 163 PSSVDSQTAMVLVNAIAFKGLWEKAFKAEDTQTIPFRVTEQESKPVQMMYQIGSFKVASM 222


protein Z1:222 DSLKV 226
                 S  K+
ovalbumin: 223 ASEKM 227
```

Fig. 3. The alignment of amino acid sequence (102nd to 226th residue) of barley protein Z1 in endosperm (EMBL Data Library entry number S25625) and that (103rd to 227th residue) of ovalbumin from Japanese quail (Mucha et al., 1990) according to the BLAST algorithm. The conserved aromatic amino acids are indicated as bold letters. +, amino acids with similar characteristics in both proteins.

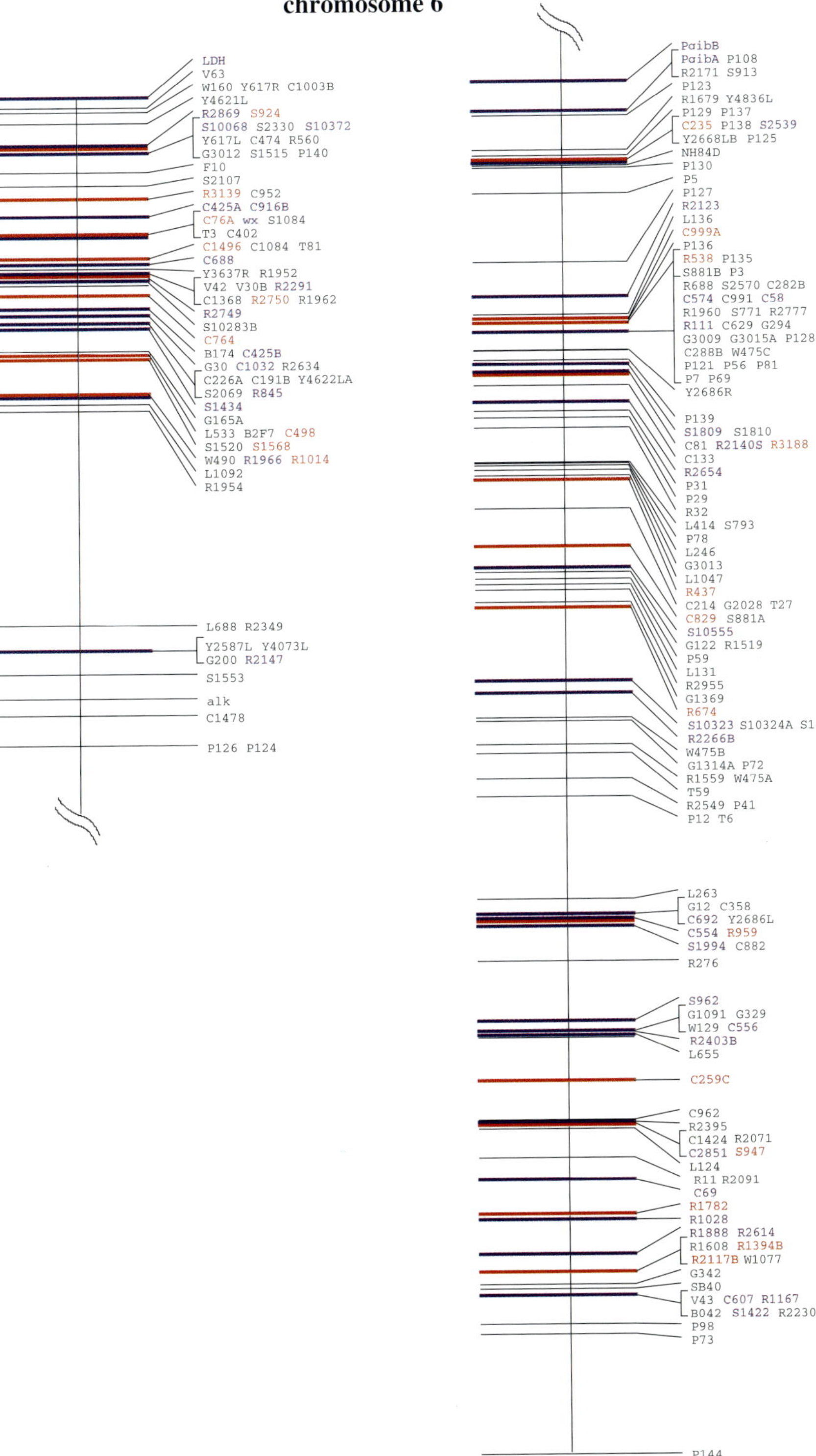

Fig. 4. The linkage map of rice chromosome 6 is shown with coloured DNA markers. Purple and red coloured bars indicate the loci of putatively identified gene names from plants and other organisms, respectively.

chromosome 6

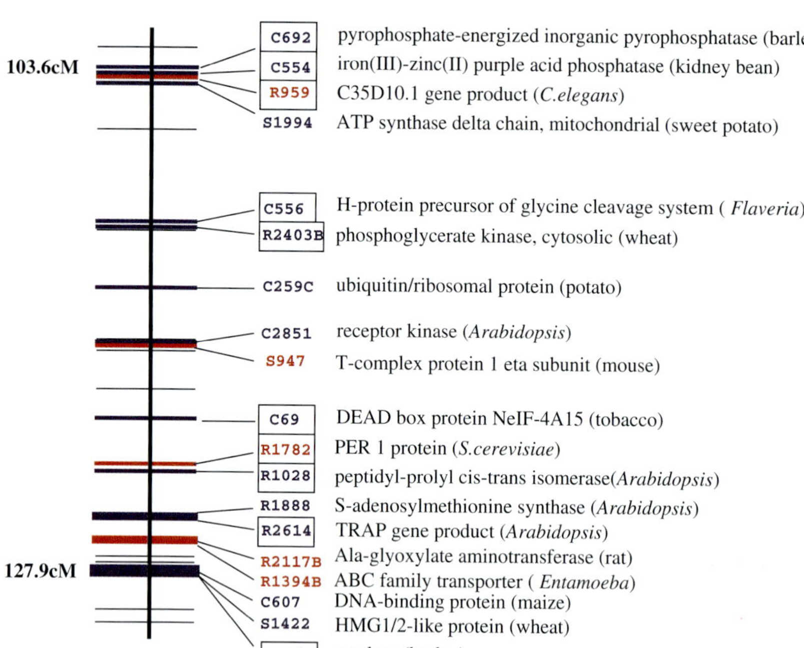

Fig. 5. The linkage map of the region from 103.6 cM to 127.9 cM from one end of chromosome 6. Purple and red coloured bars indicate the loci of putatively identified gene names from plants and other organisms, respectively. The loci within boxes were confirmed to be located on specific YAC clones.

These 279 and 291 sequences with putatively assigned protein names were ranked in order of their degree of matching to each other. From the top of the rank, twenty protein names were grouped into four categories: monocot plants, dicot plants, vertebrates and others. For example, within the twenty sequences with the highest degree of matching (100 to 95.4% match) in the 2,4-D callus library, all 20 are similar to gene products from plants. Among them, the number of gene product names originating from monocots was 13 and those from dicots was 7. On the other hand, within the twenty sequences with the lowest degree of matching (44.3 to 33.8%), only 4 of 20 are similar to gene products from plants. In the root cDNA library the same tendency was observed in the distribution of original species with similar gene products to rice root cDNAs. The distribution profiles for both cDNA libraries are shown in Figs 1 and 2.

These results strongly suggest the usefulness of amino acid sequence information from rice cDNAs for the study of genes from other monocots, such as maize and barley. Random cloning and partial sequencing of rice cDNAs is also a good way to make a cDNA catalogue for all members of the grass family. In addition, although the sequence information from both callus and root tissues gave nearly the same tendency for the distribution of targetted species, the absolute numbers for dicot plants and vertebrates were different. This might reflect a subtle difference in the profile of genes expressed in both rice tissues. Further analysis using cDNA libraries of other tissues, such as green or etiolated shoot and panicle, seems to be required to confirm this point.

MAPPING cDNAs AS RFLP PROBES

The next step for effective use of cDNAs as tools for elucidating conservative gene alignment among the *Gramineae* is the construction of a precise and detailed linkage map. In RGP, 186 F$_2$ progeny from a cross between *japonica* variety, Nipponbare, and *indica* variety, Kasalath, were used to make a linkage map. RFLPs were mainly detected by using rice cDNAs. Once the loci of cDNAs are determined, this information is indispensable for identifying the location of expressed genes. In addition, as mentioned above, the function of rice gene products is predictable with a high accuracy based on a similarity to other grass species. This makes it feasible to elucidate syntenic genome structures among grasses. RGP has published a linkage map (Kurata et al., 1994a), and is still continuing efforts to increase the number of RFLP markers on it. The latest map includes 1,904 loci of total length 1,556.0 cM and, among them, 1,300 loci are determined by cDNA clones with partial nucleotide sequence information (unpublished data). To increase this information to assign gene names, other algorithms in addition to FASTA and an increased number of data in public databases are being used.

```
Query:    77 NTKVILNGGQRVTFARSDGYFAFHNVPXGTHLIEVSSLGYLFSPVRVDISSR 128
             + ++ LN GQ +  F R D  F    VP GT+++++ +   ++F P+RVDI+S+
Sbjct:   172 SARIHLNHGQYMGFVRQDCTFRVDFVPTGTYIVQIENTDFVFEPIRVDITSK 223
```

Fig. 6. The alignment of amino acid sequence (77th to 128th) of rice cDNA R959 as a query and that (172nd to 223rd) of C35D10.1 gene product of *C. elegans* as a subject, according to the BLAST algorithm. The conserved aromatic amino acids are indicated as bold letters. +, amino acids with similar characteristics in both proteins.

RICE LINKAGE MAP WITH ASSIGNED GENE NAMES

To accomplish the requirement described above, the BLASTP algorithm (Altschul et al., 1990) was used against a non-redundant protein sequence database which is made of PIR, SWISS-PROT, PRF and translated protein sequences from GenBank. Although the score and Poisson probability defined by BLASTP were the primary points for the judgement of similarity between query and subject sequences, the degree of conservation of aromatic amino acids, such as Trp, Tyr and Phe were also taken into account. As suggested by Doolittle (1983), aromatic amino acid residues are thought to reside in the interior of protein molecules and play an important role in the maintenance of protein core structure, and therefore should be hard to replace. Especially in the case of superfamily proteins (proteins distributed among a wide range of species and with different functions but similar amino acid sequences), the conservation of these amino acids might be the key to the judgement of similarity. In Fig. 3, an example of the alignment of superfamily proteins from barley and mouse are shown.

In Fig. 4, the linkage map of rice chromosome 6 is shown with putative gene function assignments indicated. Among 125 loci determined by RFLP using cDNAs as probes, 64 are identified for their gene function by a BLASTP search against the non-redundant protein sequence database. The scores for the assignment of gene function range from 54 (rice cDNA R2123 to vetch cystein proteinase) to 581 (rice cDNA R1167 to barley catalase), partly depending on the length of amino acid sequence showing similarity and partly on the frequency of appearance of each paired amino acid. The degree of conservation of aromatic and cystein residues is also taken into consideration in cases of a score of less than 100. Further, among 64 cDNAs of assigned gene function, 43 genes with similar functions to those of rice are derived from plants.

In Fig. 5, the region between 103.6 and 127.9 cM from one end of the linkage map of chromosome 6 is shown in more detail, with the gene names. Within this region, 30 cDNAs have been mapped and 19 of them were assigned gene product names. Among them, 15 rice homologues are from plants and 4 are from species other than plants. For example, rice cDNA from root, R959, showed a significant similarity to a gene product, C35D10.1, from *Caenorhabditis elegans* (GenBank accession number U21324), which was obtained by the Nematode Genome Project (Waterston and Sulston, 1995). In Fig. 6, the alignment of both sequences is shown. Within this region, 4 of 6 aromatic amino acids in R959 are conserved in the gene product of nematode. A similar sequence to R959 is also found in cDNA clones from etiolated shoot and callus cultured in the presence of 2,4-D or naphthaleneacetic acid. Although it is now unclear whether these gene products from rice and nematode have the same function or not, the existence of the protein family shown in Fig. 6 is unambiguous.

Among the *Gramineae*, the order of genes along chromosomes is now known to be conserved. Because of its smallest genome size among the *Gramineae*, rice was chosen as a basic species for the synteny analysis. Until now, data on synteny between rice and wheat (Kurata et al., 1994a) and rice and maize (Ahn and Tanksley, 1993) have been published and synteny analyses between rice and other grasses, such as foxtail millet and barley, are underway. To effectively exchange locus information among grass maps, loci with known gene names are indispensable. As shown in Fig. 5, the loci of barley homologues of pyrophosphate-energized inorganic pyrophosphate and catalase and wheat homologues of phosphoglycerate kinase and HMG1/2-like protein were clarified by similarity search. If the genes corresponding to them were mapped on their own linkage map, a syntenic region among rice, barley and wheat genomes involving these genes should become clear.

In Fig. 5, four gene names originally known from *Arabidopsis* were found. As for *Arabidopsis*, catalogueing of cDNAs is now going on in France (Hofte et al., 1993) and the US (Newman et al., 1994). Until now, the existence of syntenic gene ordering between monocots and dicots was unknown. However, by comparing nucleotide sequences, the number of homologues between rice and *Arabidopsis* is expected to increase. In addition, many genes found in plants, such as those shown in Fig. 5, are thought to exist in both rice and *Arabidopsis*. Further mapping information based on a physical map should be more reliable and promising. The screening of a rice YAC library (Umehara et al., 1995) by RFLP probes resulted in the determination of the physical position of genes with putatively assigned names. By combining such maps of rice and *Arabidopsis*, collinearity of genes between monocots and dicots could be elucidated.

This work was supported by the Ministry of Agriculture, Forestry and Fisheries of Japan (MAFF) and the Japan Racing Association (JRA).

REFERENCES

Ahn, S. and Tanksley, S. D. (1993). Comparative linkage maps of the rice and maize genome. *Proc. Nat. Acad. Sci. USA* **90**, 7980-7984.

Altschul, S. F., Gish, W., Miller W., Myers E. W. and Lipman D. J. (1990). Basic local alignment search tool. *J. Mol. Biol.* **215**, 403-410.

Doolittle, R. F. (1983). Angiotensinogen is related to the antitrypsin-antithrombin-ovalbumin family. *Science* **222**, 417-419.

Hofte, H., Desprez, T., Amselem, J., Chiapello, H., Caboche, M., Mopisan, A., Jourjon, M.-F., Charpenteau, J.-L., Berthomieu, P., Guerrier, D. et al. (1993). An inventory of 1152 expressed sequence tags obtained by partial sequencing of cDNAs from Arabidopsis thaliana. *Plant J.* **4**, 1051-1061.

Kurata, N., Moore, G., Nagamura, Y., Foote, T., Yano, M., Minobe, Y. and Gale, M. (1994a). Conservation of genome structure between rice and wheat. *Bio/technology* **12**, 276-278.

Kurata, N., Nagamura, Y., Yamamoto, K., Harushima, Y., Sue, N., Wu, J., Antonio, B. A., Shomura, A., Shimizu, T., Lin, S.-Y. et al. (1994b). A 300 kilobase interval genetic map of rice including 883 expressed sequences. *Nature Genet.* **8**, 365-372.

Mucha, J., Klaudiny, J., Klaudinyova, V., Hanes, J. and Simuth, J. (1990). The sequence of Japanese quail ovalbumin cDNA. *Nucl. Acids Res.* **18**, 5553.

Newman, T., de Bruijn, F. J., Green, P., Keegstra, K., Kende, H., McIntosh, L., Ohlrogge, J., Raikhel, N., Somerville, S., Thomashow, M., Retzel, E. and Somerville, C. (1994). Genes galore: a summary of methods for accessing results from large-scale partial sequencing of anonymous Arabidopsis cDNA clones. *Plant Physiol.* **106**, 1241-1255.

Pearson, W. R. and Lipman, D. J. (1988). Improved tools for biological sequence comparison. *Proc. Nat. Acad. Sci. USA* **85**, 2444-2448.

Sasaki, T., Song, J., Koga-Ban, Y., Matsui, E., Fang, F., Higo, H., Nagasaki, H., Hori, M., Miya, M., Murayama-Kayano, E. et al. (1994). Toward cataloguing all rice genes: large-scale sequencing of randomly chosen rice cDNAs from a callus cDNA library. *Plant J.* **6**, 615-624.

Umehara, Y., Inagaki, A., Tanoue, H., Yasukochi, Y., Nagamura, Y., Saji, S., Otsuki, Y., Fujimura, T., Kurata, N. and Minobe, Y. (1995). Construction and characterization of a rice YAC library for physical mapping. *Molecular Breeding* **1**, 79-89.

Waterston, R. and Sulston, J. (1995). The genome of *Caenorhabditis elegans*. *Proc. Nat. Acad. Sci. USA* **92**, 10836-10840.

Comparative analysis of plant genome architecture

J. S. Heslop-Harrison

Department of cell Biology, John Innes Centre, Colney Lane, Norwich NR4 7UH, UK

SUMMARY

Many genes are similar in most plants and it is clear that the ordering of genes is highly conserved across wide taxonomic groupings. Repetitive DNA, consisting of sequence motifs between 2 and 10,000 base pairs long, repeated many hundreds or thousands of times in the genome, represents the majority of most plant genomes and defines some of the differences between species. Some sequences are highly conserved in many species, while other sequences show species or even chromosome specificity. Different types of sequences have markedly contrasting genomic distributions; even among tandem repeats, some are sub-terminal, some paracentromeric and others intercalary. The reasons for these different distributions are largely unknown, and mechanisms of homogenization, dispersion and amplifications are the subject of much speculation. Aspects of plant genome architecture – the organization of repetitive and single-copy DNA sequences along the chromosomes, and the positioning of those sequences within the nucleus at interphase – have important consequences for plant genetics. Models of large scale genome organization may be useful in learning the function of different components of the genome, in evolutionary studies and in plant breeding.

Key words: Genome organization, In situ hybridization, Repetitive DNA, Nuclear architecture, Comparative genomics

INTRODUCTION

Genetic maps are now available for almost all the important crop species in the world (see Schwarzacher, 1994). We can use these maps for marker assisted selection in plant breeding programmes, for targeted gene cloning, and to assist fundamental investigations of plant genome structure. It is now clear that many genes are similar in most plants (see Sasaki, 1996, or Delseny et al., 1996, both this volume), and that not only the sequence of the transcribed proteins, but also the ordering of genes is highly conserved across wide taxonomic groupings (Moore et al., 1995; Ahn and Tanksley, 1993). Our understanding of plant genes is increasing rapidly as knowledge of their DNA sequence and the function of proteins and signal molecules is brought together with information about the developmental timing and organ specificity of gene expression. Many general models of plant development, based on model and complex organisms, are now being elucidated.

Despite these similarities in the genes, plant genomes vary greatly in their size or DNA content (among the species described in this volume, from little more than 100 Mbp in *Arabidopsis thaliana* (see Delseny et al., 1996, this volume) to more than 23,000 Mbp in *Pinus elliottii* var. *elliottii* (see Doudrick, 1996, this volume), both species that are considered diploid). The bulk of the differences in the large genomes consists of repetitive DNA sequences, and even in the smaller genomes, almost half of the DNA is repetitive. There is an important requirement to understand the large scale organization of the plant genome, including not only single-copy but also repetitive sequences. Such knowledge is essential for studies ranging from evolutionary biology, through gene expression, understanding chromosome behaviour and meiosis, to facilitating gene transfer and plant breeding.

A comparative approach to the analysis of large-scale genome structure, involving examination of widely different species, is important to understand the nature of repetitive DNA sequences and their modes of evolution. Detailed knowledge of single species is useful, but will not be easily able to highlight universal features of genome structure. Comparative analyses may give indications of the importance of different types of repetitive sequences and their importance in evolution and development.

GENOMIC IN SITU HYBRIDIZATION

The use of total genomic DNA as a probe for in situ hybridization shows the extensive differences in the genomes of related species (see Schwarzacher, 1996, this volume). In hybrids between even closely related species, such as *Hordeum vulgare* and *H. bulbosum* in the same section of the genus (Anamthawat-Jónsson and Heslop-Harrison, 1993), or the yellow flowered *Crocus* species *C. flavus* and *C. angustifolius* (Orgaard et al., 1995), total genomic DNA from the parental or ancestral species can be isolated, labelled, and used for in situ hybridization to identify the chromosomes arising from the species-of-origin in the hybrid.

Hexaploid wheat lines with chromosomes substituted from related wild species and crops are of great interest for the introduction of new genetic variation including quality and disease resistance characters. Hexaploid bread wheat, *Triticum aestivum,* includes three genomes, A, D, and B, that come from the species similar to the diploids *Triticum monococcum, Aegilops squarrosa* and perhaps *Aegilops searsii*, respectively. About half of the world's wheat area, including all the high yielding varieties, includes a rye chromosome arm, 1RS, translocated to the wheat chromosome arm 1BL. Another line of interest for plant breeding includes chromosome segments from *Aegilops umbellulata,* a species with novel high molecular mass protein components that confer high strength to flour (Law et al., 1984). Genomic DNA from *Aegilops umbellulata* can be used to label the chromosomes (Fig. 1) or chromosome segments (Fig. 2) of *Ae. umbellulata*-origin in the wheat background. A second probe of a cloned repetitive DNA sequence with a characteristic genomic distribution can be used to identify the chromosomes involved in the substitution or translocation events (Castilho et al., 1996a). Most single-copy DNA clones isolated from hexaploid wheat and used for RFLP analysis also hybridize to *Ae. umbellulata* (Castilho, 1995; Castilho et al., 1996b), indicating that the repetitive DNA content, represented in the genomic DNA probe, is much more widely diverged between the species.

REPETITIVE DNA SEQUENCES

Genomic in situ hybridization is valuable to discriminate genomes in hybrids, but we are also interested in the repetitive sequences themselves, to examine and compare their evolution and distribution.

Various classifications of repetitive sequences can be made, one being the length of the repeated sequence motif. The motif length varies from simple sequence repeats of dinucleotides, such as GA, through tri-, tetra- and hexa-nucleotides, to nucleosome repeats of about 180 base pairs (the length of DNA coiling around a single nucleosome), up to 10,000 or more base pairs, for example in the rDNA, a repeated unit including three genes, for 18S, 5.8S and 25S rRNA genes, and the transcribed and non-transcribed spacers.

While a few highly repetitive sequences have important functions – the telomeric repeat and rDNA genes in particular – most have no known role. Some are tandemly repeated, and normally particular sequence classes are located on many chromosomes. However, detailed analysis by a combination of sequencing, and use as primers in situ on chromosomes and in the polymerase chain reactions with genomic DNA templates shows that there are variants specific to only one or a few chromosomes in some animals (Bernard and Wood, 1993; Kipling et al., 1994) and perhaps plants (unpublished).

Another important classification of repetitive DNA is its species specificity (Anamthawat-Jónsson and Heslop-Harrison, 1992; Metzlaff et al., 1986; Rayburn and Gill, 1986; Gazdova et al., 1995). Some sequences are found in only a single species or even ecotype, or a small group of species within a tribe, while other sequences have very wide distribution in many members of a family. Sequences may show variation at the sequence or copy number levels between and sometimes within different species. Such distinctions may be of considerable significance in taxonomy. How uniform are the sequences of repeat units, and how variable is their genomic organization?

Repetitive sequences can also be classified by their genomic locations: many are tandemly repeated in blocks of many copies, but some are dispersed throughout the genome. In situ hybridization of different, tandemly repeated DNA sequences which comprise between 1 and 10% of a genome, shows that some sequences are centromeric (Figs 3, 4), other subtelomeric (Fig. 1), and others with different but characteristic locations at intercalary sites on all or most chromosome arms. The reasons why the sequences are highly amplified at characteristic genomic positions are unknown, but of great interest. As yet there are no features of the primary sequence that would enable one to predict the location of a sequence motif.

rRNA genes

The physical distribution of the highly conserved 5S and 18S-5.8S-25S rRNA gene loci on the chromosomes are useful markers to examine genome evolution, and have been widely

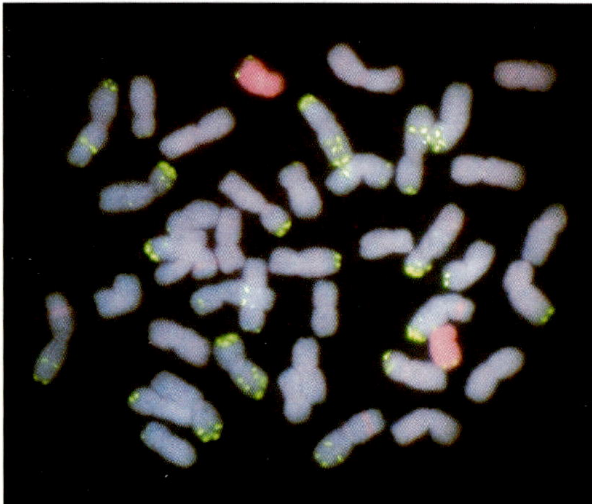

Fig. 1. A metaphase of wheat with a telocentric chromosome pair from *Aegilops umbellulata (*labelled red) substituting for one wheat chromosome pair following DNA:DNA in situ hybridization with labelled genomic *Ae. umbellulata* DNA. Wheat chromosomes show blue DAPI staining, and an additional repetitive DNA probe, detected green, enables many chromosomes to be identified. (See Castilho et al., 1996.)

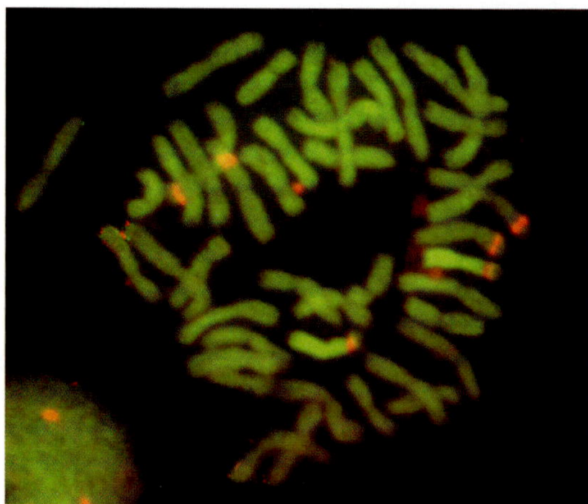

Fig. 2. A metaphase of a wheat with two *Ae. umbellulata*-wheat recombinant chromosomes (bright green, centre right and lower centre). rDNA sites on nucleolar organizing chromosomes are labelled red; a dull terminal chromosome segment of wheat origin is visible distal to the *Ae. umbellulata*-origin rDNA sites.

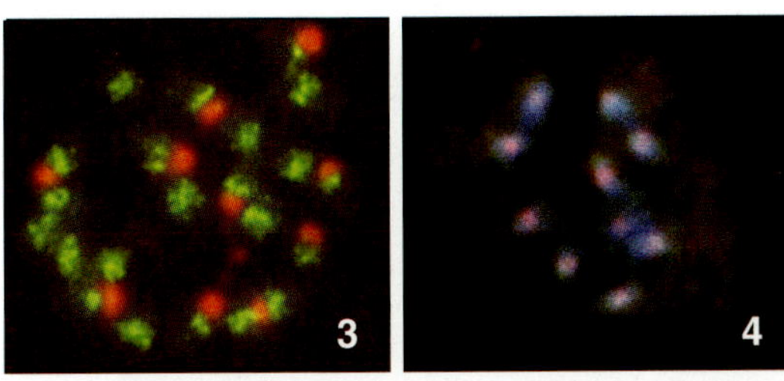

Fig. 3. A metaphase of *Vigna unguiculata* showing hybridization of one probe to the centromeric region of all 22 chromosomes (green), and rDNA sites on 10 chromosomes (red). (See Galasso et al., 1995.)
Fig. 4. A metaphase of *Arabidopsis thaliana* showing a repetitive DNA sequence at the centromeres of all 10 chromosomes. (See Heslop-Harrison and Maluszynska, 1994.)

examined extensively in several species. Within the Triticeae, the 18S-25S and 5S rDNA sequences show different rearrangements and deletions in barley, rye, wheat and their wild relatives, enabling detection of genome rearrangements (Castilho and Heslop-Harrison, 1995). Importantly, analysis of the presence and order of the two gene clusters show that similar events have occurred more than once during the evolution of the *Aegilops* and *Triticum* genera (see Fig. 2). In *Vigna* (Fig. 3), we have found polymorphisms in 18S-25S rDNA site numbers between different accessions of cowpea and its wild relatives in the same species (Galasso et al., 1995, and in preparation), again showing the rapid evolution of these sequences, and their value as chromosome markers in breeding programmes involving both inter- and intra-specific hybridization.

Telomeric sequences

The telomeric repeat, in plants similar to the septamer (GGGATTT)n, is synthesized by a telomerase, an enzyme which includes a nucleic acid template and adds the sequence to the ends of chromosomes. Endpoints, representing telomere locations, are needed on any complete map of a species, and can be mapped by pulse-field gel electrophoresis (PFGE) (Broun et al., 1992; Ganal et al., 1991).

Sequences homologous to the telomeric repeat may be localized largely at the telomeres (in rye), near some centromeres and telomeres (e.g. in *Arabidopsis thaliana)*, or with major intercalary sites in the middle of chromosome arms, and minor sites at telomeres (e.g. in pines; see Doudrick, 1996, this volume).

In rye, we have examined the organization of the authentic telomeric sequence (TTTAGGG)n and other proximal repetitive DNA sequences by a combination of in situ hybridization and Southern hybridization to pulse-field gel electrophoresis (PFGE) separated DNA (Vershinin et al., 1995). Many blocks show a complex organization, and some show high levels of instability. In many species, these sequences are located near the nuclear envelope within interphase nuclei, and may have characteristic clustering.

Centromeric sequences

Many plants have clusters of highly repetitive sequences at or near their centromeres. These sequences often stain characteristically, so both *Arabidopsis* and pines (see Doudrick, 1996, this volume) show centromeric regions that stain brightly with the fluorochrome DAPI. Other species show no such characteristic bands at their centromeres, which normally represent tandem repeats of a highly repetitive DNA sequences.

Two different repetitive sequences, approximately the length of one nucleosome, have been characterized in the tetraploid species *Brassica napus* (oilseed rape) and the diploid ancestors species *B. campestris* and *B. oleracea* (cabbage) (Harrison and Heslop-Harrison, 1995; Chevre et al., 1994). In situ hybridization shows that the two sequences are located near the centromeres of many, but not all, chromosome pairs in both diploid species. Reduction of stringency of hybridization gave centromeric hybridization sites on more chromosomes, indicating that there are divergent sequences present on other chromosomes. In the tetraploid species derived from the diploids, the number of hybridization sites was different from the sum of the diploid ancestors, and some chromosomes had both sequences, indicating relatively rapid homogenization and copy number evolution since the origin of the tetraploid species (Harrison and Heslop-Harrison, 1995). In other species in the Crucifers, changes that have occurred during polypolidization and formation of new species can also be examined by looking at the diversification and distribution of centromeric DNA sequences.

How uniform are the sequences of repeat units (between related species and within one species) and how variable is their genomic organization? Such distinctions may be of considerable significance in taxonomy. For example, the paracentromeric DNA sequences comprising 2-3% of the genome of the species widely known as *Cardaminopsis arenosa* (L.) Hayck (in the tribe Arabideae) are extremely similar to those from *Arabidopsis thaliana* L. (in the tribe Sysimbrideae; Fig. 4). Thus, the uncommon alternative classification of *C. arenosa* as *A. arenosa* (L.) Lawalrée is much more likely to be correct (Kamm et al., 1995).

Within species, there are also data enabling comparison of the primary sequence and chromosomal distribution of different variants of the same DNA family within species. In the wild beet species *Beta procumbens,* members of a sequence family pTS have been analysed, and show between 80 and 99% homology, with variants involving both substitution and deletion of nucleotides (Schmidt and Heslop-Harrison, 1996a). Fig. 5 shows in situ hybridization of two members of this repeat family to a pachytene chromosome preparation of *Beta procumbens*. Both sequences are localized near the centromeres of all nine chromosome pairs, but they are not entirely co-located. Some regions consist largely of the variant labelled red, while others are labelled with the green variant. In one chromosome, green variants flank the red variant. Fig. 6 shows two different repetitive elements with contrasting genomic distributions hybridizing to sugar beet chromosomes. Both show some site-specific amplification, although one is present throughout most of the genome (Fig. 6b), while the other is concentrated at some centromeric sites (Fig. 6c).

These data indicate that there are mechanisms allowing sequences to distribute between chromosomes and to amplify, but that homogenization of the sequences within and between chromosomes is not complete. However, preliminary results indicate that homogenization mechanisms for repetitive sequences may be more active in plants than in mammalian chromosomes (Schwarzacher et al., 1996).

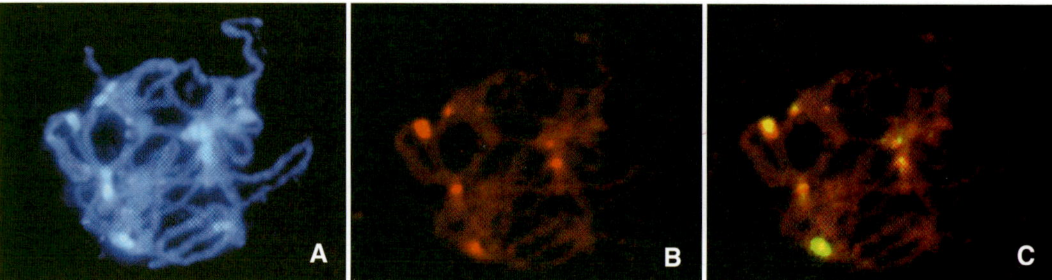

Fig. 5. A pachytene nucleus of the wild beet species *Beta procumbens* after (a) DAPI staining; (b) in situ hybridization of one DNA probe (red) and (c) with a second probe with high homology to the probe in b (yellow) showing the close but not entirely overlapping localization of the two probes. (See Schmidt and Heslop-Harrison, 1996a).

Retroelements

Many tandem repeats are presumably amplified, and perhaps homogenized, by unequal crossing over. But the transposition mechanisms by which most repetitive sequences move from one chromosome to another without major genomic reorganizations, are largely unknown, and lead to characteristic genomic locations along chromosomes and at interphase. Retrotransposons include genes for reverse transcriptase, a coding sequence which is related to their mode of amplification and dispersion within the genome. In sugar beet, we have isolated members of both the major classes of retrotransposons (the LTR, Ty-copia type and the non-LTR, LINElement type), and can show that they are significant components of the genome with strikingly different genomic organizations (Schmidt et al., 1995).

Microsatellites

Microsatellites, consisting of tandem repeats of dinucleotide to tetranucleotide motifs, are ubiquitous in plant genomes. Length variation in microsatellite arrays is rapidly becoming a widely used and robust method for genetic mapping; sequences flanking a microsatellite repeat array are conserved and can be used to amplify the intervening microsatellite by PCR. Since the array lengths are somatically stable and inherited in co-dominant Mendelian manner but highly polymorphic between lines, they can be used as segregating genetic markers (Wu and Tanksley, 1993; Roder et al., 1995; Saghai Maroof et al., 1994; Bell and Ecker, 1994) in many species.

Microsatellites can be used directly as probes for assessment of genetic variability by Southern hybridization (Zischler et al.,

1991; Zhao and Kochert, 1993; Schmidt et al., 1993). Some simple sequence repeats have been located by in situ hybridization and show specific banding patterns (Schmidt and Heslop-Harrison, 1996b) and characteristic chromosomal distributions (Pedersen and Linde-Laursen, 1994). Fig. 7 shows hybridization of the microsatellite probe (GATA)4 to chromosomes of sugar beet. This shows a dispersed distribution over the lengths of all 18 chromosomes, with large clustered sites near the centromeres of six chromosomes. Other microsatellite motifs show exclusion or amplification at centromeric locations.

Sequence methylation

The extent of methylation of the repetitive sequences is variable. While many repetitive DNA sequences from plants show methylation of most cytosine residues, some repetitive DNA may be under-methylated, for example in *Pennisetum glaucum* (Kamm et al., 1994). In the case of the ribosomal genes, reduction of methylation may be associated with expression of the genes, and may be modulated during development (e.g. Neves et al., 1995). Methylation may also be important for centromere activity and the stability of chromosomes during division.

NUCLEAR ARCHITECTURE

The organization of different types of DNA sequences along chromosomes is one aspect of plant genome architecture. A second aspect is the three-dimensional arrangement of the

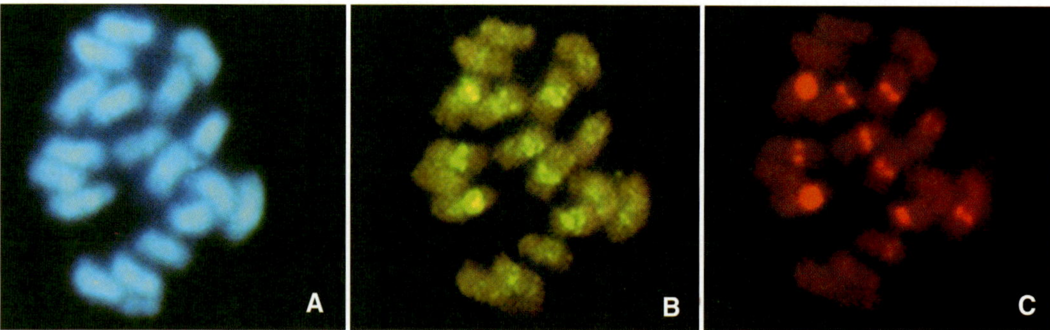

Fig. 6. The distribution of two different repetitive DNA sequences on metaphase chromosomes of *Beta vulgaris*. (a) DAPI stained chromosomes; (b) Hybridization of a dispersed repetitive element over all 18 chromosomes; (c) a tandemly repeated sequence hybridizing strongly to the centromeres of most but not all chromosomes. (Schmidt et al., unpublished.)

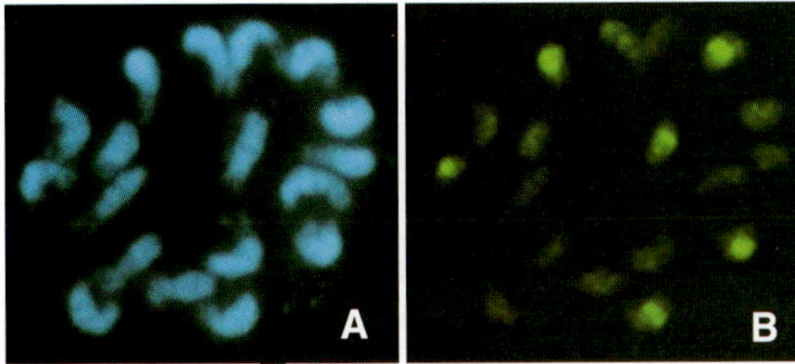

Fig. 7. (a) DAPI staining showing the chromosomes of *Beta vulgaris* (2n=18). (b) In situ hybridization of the synthetic microsatellite oligonucleotide (GATA)4 shows major sites of hybridization near the centromeres of six chromosomes and diffuse hybridization along the lengths of all 18 chromosomes. (See Schmidt and Heslop-Harrison, 1996b.)

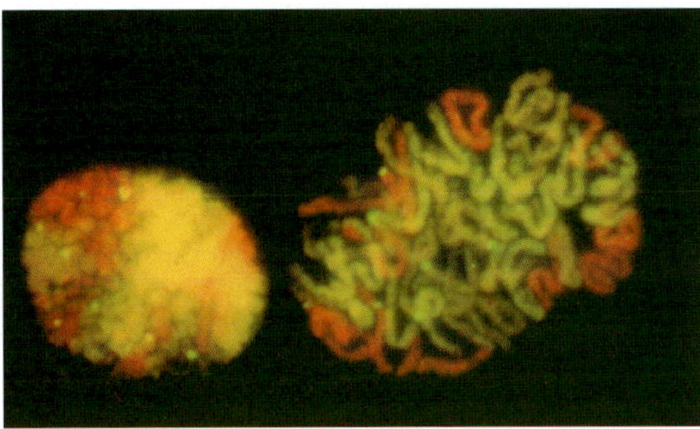

Fig. 9. As Fig. 8, with wheat-origin chromosomes labelled yellow (A genome) or brown (B genome) and wheatgrass-origin chromosomes red. In two prophase nuclei, the chromosomes or domains occupied by the three genomes can be distinguished by their colour, and show limited intermixing of chromosomes. (See Kosina and Heslop-Harrison, 1996.)

sequences within the nucleus (Heslop-Harrison and Bennett, 1990), particularly at interphase when the genes are actively expressed and DNA is being replicated. The cell nucleus is highly organized, with large amounts of DNA and the complex machinery associated with the regulation and copying of DNA is packed into a small volume. However, relatively little is known about nuclear organization, even at the descriptive level, in plants (Heslop-Harrison and Bennett, 1990) or in animals (Cremer et al., 1993), although there is little doubt that the arrangement is of significance for many aspects of plant genetics including gene regulation, recombination and some abnormalities of development. Repetitive DNA plays a role in plant genome organization, and is undoubtedly important in nuclear behaviour and evolution. Its roles in gene expression, meiosis (see Schwarzacher, 1996, this volume) and other aspects of plant genetics are largely unexplored.

As well as giving information about the relationships and evolution of genomes, and proving valuable to follow alien chromosomes and chromosome segments in plant breeding programmes, genomic in situ hybridization is also valuable for understanding aspects of nuclear architecture at interphase. Figs 8 and 9 show nuclei from hybrids between a tetraploid wheat, with the genomes A and B, and wheatgrasses, with another genome (Kosina and Heslop-Harrison, 1996). Discrimination of the wheat and wheatgrass chromosomes is clear

after probing with genomic DNA from wheatgrass and *Triticum monococcum,* and shows that the wheatgrass chromosomes occupy defined domains within the interphase nucleus, and are not entirely intermixed. Furthermore, Fig. 9 shows some discrimination between wheat A and B genome chromosomes (more yellow and more brown, respectively) is possible at both metaphase and interphase, indicating that chromosomes from these genomes too lie in distinct domains.

Knowledge of genome architecture is likely to be helpful to targeted approaches to genetic manipulation. The genomic context of a sequence is unlikely to be entirely random; if so, evolutionary changes in chromosome number and structure would probably be much more frequent. However, we have little knowledge of why repetitive sequences do things: is the three-dimensional architecture of the nucleus important for function? Do repetitive sequences put coding or regulatory sequences in particular nuclear positions? Why are different sequences located in particular sites in the genome?

CONCLUSIONS

The data discussed above indicate that knowledge of more than DNA sequence, chromosome behaviour or genetic markers is necessary to build up a complete picture of the genome. Repetitive DNA sequences are an important component of the plant genome and play a role in both its structure and activity. Knowledge of genome architecture is already useful to provide markers in plant breeding, particularly through genomic in situ hybridization, and is starting to be useful to see the types of changes breeders might make to a genome. In the future, the data are likely to be helpful to targeted approaches to genetic manipulation. The genomic context of a sequence is unlikely to be entirely random – if so, evolutionary changes in chromosome number and structure would probably be much more frequent. However, we have little knowledge of why repetitive sequences do things: is the three-dimensional architecture of the nucleus important for function? Do repetitive sequences put coding or regulatory sequences in particular nuclear positions? Why are different sequences located in particular sites in the genome? Examination of the changes that have occurred during long periods of plant evolution indicate the types of

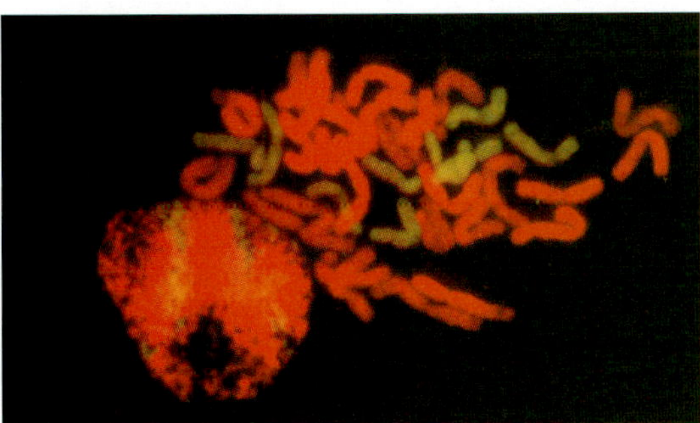

Fig. 8. In situ hybridization to durum wheat (originally 28 chromosomes, AABB genomes) × wheatgrass (originally 14 chromosomes, here with 9 chromosomes remaining labelled yellow) amphiploid hybrid nuclei. A (bright red, e.g. lower chromosome, far right) and B genome (dull red, e.g. upper chromosome, far right) wheat-origin chromosomes are distinguished at metaphase. At interphase, the wheatgrass-origin chromosome occupy two distinct domains and are not intermixed with the wheat-origin chromosomes.

changes that plant breeders will be able to make in plant genomes during selection.

Blocks of repetitive DNA may separate genetically closely linked markers. Knowledge of both the genetic and physical map is essential for discovering the feasibility of 'gene walking' and other gene isolation strategies (Heslop-Harrison et al., 1993). There are also clear correlations between nuclear architecture and gene location within the nucleus.

Work on synteny and the unification of plant genomes is showing that not only technical information, but makers, genes and the controlling sequences can all be transferred directly between species. How much this applies to repetitive DNA is unknown. The rapidly accumulating results from projects on the model Crucifer *Arabidopsis thaliana* are providing insights into gene action and expression that could only be dreamt of three years ago. Clearly, some knowledge from *A. thaliana* can be applied to *B. napus,* although the actual limits of cross-application of knowledge are far from being known. It is probable that knowledge of chromosome structure and behaviour will assist in finding the limitations.

As well as a gap between cytogenetic and genetic information, there is often a significant gap between the detailed knowledge of the genome and the application of that knowledge to plant breeding (Heslop-Harrison, 1993; Heslop-Harrison and Schwarzacher, 1993). How useful is fundamental knowledge to applied problems? Will lines with, say, translocations be so poor that they should be discarded with the minimum of effort? Or are valuable recombinants lost in this process? Will particular useful combinations of features be much more difficult to combine unless the physical structures of the genomes are known in some detail?

We are now constructing a comprehensive and quantitative model of variable and constant parts of the plant genome. Such models of large scale genome organization may be useful in learning the function of different components of the genome, and in evolutionary studies. Since repetitive DNA changes are frequent, perhaps we can learn more about what manipulations are possible in plant genomes by examining the changes already made between related species during evolution or already during plant breeding programmes.

I am grateful to my many collaborators, cited in references, for their immense contribution to this programme and permission to use their photographs. I thank BBSRC for support of the work described here under a Plant and Animal Genome Analysis grant.

REFERENCES

Ahn, S. and Tanksley, S. D. (1993). Comparative linkage maps of the rice and maize genomes. *Proc. Nat. Acad. Sci. USA* **90**, 7980-7984.

Anamthawat-Jónsson, K. and Heslop-Harrison, J. S. (1992). Species specific DNA sequences in the Triticeae. *Hereditas* **116**, 49-54.

Anamthawat-Jónsson, K. and Heslop-Harrison, J. S. (1993). Isolation and characterization of genome-specific DNA sequences in Triticeae species. *Mol. Gen. Genet.* **240**, 151-158.

Bell, C. J. and Ecker, J. R. (1994). Assignment of 30 microsatellite loci to the linkage mmap of Arabidopsis. *Genomics* **19**, 137-144.

Bernard, L. E. and Wood, S. (1993). Human chromosome 5 sequence primer amplifies Alu polymorphisms on chromosomes 2 and 17. *Genome* **36**, 302-309.

Broun, P., Ganal, M. W. and Tanksley, S. D. (1992). Telomeric arrays display high levels of heritable polymorphism among closely related plant varieties. *Proc. Nat. Acad. Sci. USA* **89**, 1354-1357.

Castilho, A. (1995). Molecular cytogenetic analysis of recombinant chromosomes in wheat-*Aegilops umbellulata* lines. PhD thesis, University of East Anglia, Norwich.

Castilho, A. and Heslop-Harrison, J. S. (1995). Physical mapping of 5S and 18S-25S rDNA and repetitive DNA sequences in *Aegilops umbellulata. Genome* **38**, 91-96.

Castilho, A., Miller, T. E. and Heslop-Harrison, J. S. (1996a). Physical mapping of sub-centromeric and intercalary translocation breakpoints in a set of wheat-A*egilops umbellulata* recombinant lines using *in situ* hybridization. *Theor. Appl. Genet.* (in press).

Castilho, A., Miller, T. E. and Heslop-Harrison, J. S. (1996b). Analysis of a set of homoeologous group 1 wheat-*Aegilops umbellulata* recombinant chromosome lines using genetic markers. *Theor. Appl. Genet.* (in press).

Chevre, A. M., Eber, F., Margale, E., Kerlan, M. C., Primard, C., Vedel, F., Delseny, M. and Pelletier, G. (1994). Comparison of somatic and sexual Brassica napus – Sinapis alba hybrids and their progeny by cytogenetic studies and molecular characterization. *Genome* **37**, 367-374.

Cremer, T., Kurz, A., Zirbel, R., Dietzel, S., Rinke, B., Schrock, E., Speicher, M. R., Mathieu, U., Jauch, A., Emmerich, P., Scherthan, H., Ried, T., Cremer, C. and Lichter, P. (1993). Role of chromosome territories in the functional compartmentalization of the cell nucleus. *CSH Symp. Quant. Biol.* **58**, 777-791.

Delseny, M., Raynal, M., Laudie, M., Varoquaux, F., Comella, P., Wu, H. J., Cooke, R. and Grellet, F. (1996). Sequencing and mapping the *Arabidopsis* genome: a weed model for real crops. *SEB Symp.* **50**, 5-9.

Doudrick, R. L. (1996). Genetic recombinational and physical linkage analyses on slash pine. *SEB Symp.* **50**, 53-60.

Galasso, I., Schmidt, T., Pignone, D. and Heslop-Harrison, J. S. (1995). The molecular cytogenetics of *Vigna unguiculata* (L.) Walp: the physical organization and characterization of 18S-5.8S-25S rRNA genes, 5S rRNA genes, telomere-like sequences, and a family of centromeric repetitive DNA sequences. *Theor. Appl. Genet.* **91**, 928-935.

Ganal, M. W., Lapitan, N. L. V. and Tanksley, S. D. (1991). Macrostructure of the tomato telomeres. *Plant Cell* **3**, 87-94.

Gazdova, B., Siroky, J., Fajkus, J., Brzobohaty, B., Kenton, A., Parokonny, A., Heslop-Harrison, J. S., Palme, K. and Bezdek, M. (1995). Characterization of a new family of tobacco highly repetitive DNA, GRS, specific for the *N. tomentosiformis* genomic component. *Chromosome Res.* **3**, 245-254.

Harrison, G. E. and Heslop-Harrison, J. S. (1995). Centromeric repetitive DNA in the genus *Brassica. Theor. Appl. Genet.* **90**, 157-165.

Heslop-Harrison, J. S. and Bennett, M. D. (1990). Nuclear architecture in plants. *Trends Genet.* **6**, 401-405.

Heslop-Harrison, J. S. (1993). Applications of molecular cytogenetics in the Triticeae. In *Biodiversity and Wheat Improvement* (ed. A. Damania), pp. 31-38. Chichester: Wiley and Sons.

Heslop-Harrison, J. S., Leitch, A. R. and Schwarzacher, T. (1993). The physical organization of interphase nuclei. In *The Chromosome* (ed. J. S. Heslop-Harrison and R. B. Flavell), pp. 221-232. Oxford: BIOS.

Heslop-Harrison, J. S. and Schwarzacher, T. (1993). Molecular cytogenetics – biology and applications in plant breeding. *Chrom. Today* **11**, 191-198.

Heslop-Harrison, J. S. and Maluszynska, J. (1994). The molecular cytogenetics of *Arabidopsis*. In Arabidopsis (ed. E. M. Meyerowitz and C. R. Sommerville), pp. 63-87. Cold Spring Harbor Laboratory Press, Cold Spring Harbor, NY.

Kamm, A., Schmidt, T. and Heslop-Harrison, J. S. (1994). Molecular and physical organization of highly repetitive undermethylated DNA from *Pennisetum glaucum. Mol. Gen. Genet.* **244**, 420-425.

Kamm, A., Galasso, I., Schmidt, T. and Heslop-Harrison, J. S. (1995). Analysis of a repetitive DNA family from *Arabidopsis arenosa* and relationships between *Arabidopsis* species. *Plant. Mol. Biol.* **27**, 853-862.

Kosina, R. and Heslop-Harrison, J. S. (1996). Molecular cytogenetics of an amphiploid trigeneric hybrid between *Triticum durum, Thinopyrum distichum* and *Lophopyrum elongatum. Ann. Bot.* (in press).

Kipling, D., Wilson, H. E., Mitchell, A. R., Taylor, B. A. and Cooke, H. J. (1994). Mouse centromere mapping using oligonucleotide probes that detect variants of the minor satellite. *Chromosoma* **103**, 46-55.

Law, C. N., Payne, P. I., Worland, A. J., Miller, T. E., Harris, P. A., Snape, J. W. and Reader, S. M. (1984). Studies of genetical variation affecting grain protein type and amount in wheat. In *Cereal Grain Protein Improvement IAEA*, pp. 279-300. Vienna.

Metzlaff, M., Troebner, W., Baldauf, F., Schlegel, R. and Cullum, J. (1986). Wheat specific repetitive DNA sequences – construction and characterization of four different genomic clones. *Theor. Appl. Genet.* **72**, 207-210.

Moore, G., Devos, K. M., Wang, Z. and Gale, M. D. (1995). Grasses, line up and form a circle. *Curr. Biol.* **7**, 737-739.

Neves, N., Heslop-Harrison, J. S. and Viegas, W. (1995). rRNA gene activity and control of expression mediated by methylation and imprinting during embryo development in wheat × rye hybrids. *Theor. Appl. Genet.* **91**, 529-533.

Orgaard, M., Jacobsen, N. and Heslop-Harrison, J. S. (1995). The hybrid origin of two cultivars of *Crocus* (Iridaceae) analysed by molecular cytogenetics including genomic Southern and *in situ* hybridization. *Ann. Bot.* **76**, 253-262.

Pedersen, C. and Linde-Laursen, I. (1994). Chromosomal locations of four minor rDNA loci and a marker microsatellite sequence in barley. *Chrom. Res.* **2**, 65-71.

Rayburn, A. L. and Gill, B. S. (1986). Isolation of a D-genome specific repeated DNA sequence from Aegilops squarrosa. *Plant Mol. Biol. Rep.* **4**, 102-109.

Roder, M. S., Plaschke, J., Konig, S. U., Borner, A., Sorrells, M. E., Tanksley, S. D. and Ganal, M. W. (1995). Abundance, variability and chromosomal location of microsatellites in wheat. *Mol. Gen. Genet.* **246**, 327-333.

Saghai Maroof, M. A., Biyashev, R. M., Yang, G. P., Zhang, Q. and Allard, R. W. (1994). Extraordinarily polymorphic microsatellite DNA in barley: species diversity, chromosomal locations, and population dynamics. *Proc. Nat. Acad. Sci. USA* **91**, 5466-5470.

Sasaki, T. (1996). Rice cDNAs as a model for expressed genes of plants. *SEB Symp.* **50**, 11-15.

Schmidt, T., Boblenz, K., Metzlaff, M., Kaemmer, D., Weising, K. and Kahl, G. (1993). DNA fingerprinting in sugar beet (*Beta vulgaris*) – identification of double-haploid breeding lines. *Theor. Appl. Genet.* **95**, 653-657.

Schmidt, T., Kubis, S. and Heslop-Harrison, J. S. (1995). Analysis and chromosomal localization of retrotransposons in sugar beet (*Beta vulgaris* L.): LINEs and T*y1-copia*-like elements as major components of the genome. *Chrom. Res.* **3**, 335-345.

Schmidt, T. and Heslop-Harrison, J. S. (1996a). High resolution mapping of repetitive DNA by *in situ* hybridization – molecular and chromosomal features of prominent dispersed and discretely localized DNA families from the wild beet species *Beta procumbens. Plant Mol. Biol.* **30**, 1099-1114.

Schmidt, T. and Heslop-Harrison, J. S. (1996b). The physical and genomic organization of microsatellites in sugar beet. *Proc. Nat. Acad. Sci. USA* (in press).

Schwarzacher, T. (1994). Mapping in plants: progress and prospects. *Curr. Opin. Genet. Dev.* **4**, 868-874.

Schwarzacher, T. (1996). Molecular cytogenetics and investigations of cereal genome structure and meiosis. *SEB Symp.* **50**, 71-75.

Schwarzacher, T., Wang, M. L., Leitch, A. R., Miller, N., Moore, G. and Heslop-Harrison, J. S. (1996). Flow cytometric analysis of the chromosomes and stability of a wheat cell culture line. *Theor. Appl. Genet.* (in press).

Vershinin, A., Schwarzacher, T. and Heslop-Harrison, J. S. (1995). The large scale genomic organization of repetitive DNA families at the telomeres of rye chromosomes. *Plant Cell* **7**, 1823-1833.

Wu, K.-S. and Tanksley, D. (1993). Abundance, polymorphism and genetic mapping of microsatellites in rice. *Mol. Gen. Genet.* **241**, 225-235.

Zhao, Z. and Kochert, G. (1993). Phylogenetic distribution and genetic mapping of a (GGC)n microsatellite from rice (*Oryza sativa* L.). *Plant Mol. Biol.* **21**, 607-614.

Zischler, H., Hinkkanen, A. and Studer, R. (1991). Oligonucleotide fingerprinting with (CAC)5: Nonradioactive in-gel hybridization and isolation of individual hypervariable loci. *Electrophoresis* **12**, 141-146.

Printed in Great Britain © The Society for Experimental Biology 1996
SEB0025

Physical and topographical mapping among Triticeae chromosomes

Reinhold G. Herrmann, Regina Martin, Winfried Busch, Gerhard Wanner and Uwe Hohmann

Botanisches Institut der Ludwig-Maximilians-Universität München, Menzinger Straße 67, D-80638 München, Germany

SUMMARY

Three principal approaches have been used in our laboratory to analyze Triticeae genomes. (i) Synteny analysis: synteny among different Gramineae genomes was studied employing the elegant system of the *Agropyron* chromosome-induced deletion lines of wheat. Deletion mapping, predominantly of the homoeologous group 7 chromosomes, has led to the construction of a high density physical consensus map of wheat. The integration of wheat, barley and oat RFLP markers proves the colinearity between the wheat A-, B- and D-genomes, the H-genome of barley, and the E-genome of *Agropyron*. (ii) Light microscopic in situ techniques: the recent improvement of a drop technique for plant protoplasts was crucial for the sensitivity enhancement of fluorescence in situ hybridization (FISH), the efficient preparation of plant chromosomes for high resolution scanning electron microscopy, mapping of low-copy sequences, and comparative in situ hybridization. A tandemly amplified repetitive sequence element from microdissected barley chromosomes has enabled the karyotyping of Gramineae genomes in a single step. We have isolated and characterized members of this element family from other Triticeae species using PCR. The significant interspecific sequence differences were useful to identify single plant genomes, chromosomes and chromosome segments via post-hybridization washes under different stringency conditions. These sequences are also useful for simultaneous double or triple hybridization experiments in an attempt to localize new sequences on specific chromosomes or chromosome segments. The physical mapping of the *Sec-1* locus has been refined on the satellite of chromosome 1R of rye, and the syntenic locus on barley chromosome 1H was identified. (iii) Physical mapping of rDNA sequences by high resolution electron microscopy: a method was developed for in situ hybridization and signal detection using high resolution field emission scanning electron microscopy and a backscattered electron detector. Colloidal gold particles were localized on chromosome structures resembling the 30 nm fibre. An rDNA probe was located in the secondary constriction and the highly compact adjacent regions of barley chromosomes.

Key words: Synteny, Comparative genome analysis, Chromosome ultrastructure, High resolution mapping

INTRODUCTION

The ultimate goal of plant genome and chromosome analysis is the exhaustive functional and structural description of genomes, their comparison and use. The analysis of complex eukaryotic genomes has made substantial progress during the past two decades predominantly due to the development of a molecular marker system based on restriction fragment length polymorphisms (RFLPs), mini- or microsatellites, polymerase chain reaction (PCR), random amplified polymorphic DNAs (RAPDs), or arbitrary fragment length polymorphisms (AFLPs) as well as of megabase technologies (Jordan, 1988; Saiki et al., 1985). The latter have contributed significantly to integrate genetic mapping procedures (linkage analysis) and conventional plasmid- or lambda-based cloning techniques. Basically three techniques, pulse field gel electrophoresis, cloning in yeast artificial chromosomes (YACs) or bacterial artificial chromosomes (BACs), and chromosome jumping, have pioneered the handling and/or physically mapping of long coherent DNA segments.

Molecular markers provide a useful tool in genetic fingerprinting, evolutionary and pedigree studies, or in the characterization of germ plasm stocks and represent the most efficient way for constructing detailed genetic linkage maps which can be established by conventional segregation analysis of F_1 or BC_1 populations. Until now maps have been constructed from almost a hundred plant genomes and some of them are of high resolution. Molecular markers can further serve for the isolation of developmentally crucial or economically important genes, for instance, of those conferring pathogen or drought resistance or quality characters which can usually be detected by phenotype only, since their products are not known. The marker technology also offers the unique opportunity to study quantitative, polygenic traits (QTLs; e.g. Osborn et al., 1987; Paterson et al., 1988), that is, to dissect complex developmental or metabolic processes into individual Mendelian loci and to determine the relative contribution of each locus even independent of the environment. During the last few years, comparative marker analysis has uncovered synteny, for instance among Solanaceae or Gramineae genomes, respectively, and has been extensively used to analyze plant genomes and chromosomes. Significant conservation with regard to gene content and gene order has been found among related crop species, between tomato and potato (Bonierbale et al., 1988; Tanksley et al., 1992), maize and sorghum (Hulbert et al., 1990; Whitkus et al., 1992; Berhan et al., 1993), as well as between rice, maize and wheat (Ahn and Tanksley, 1993; Ahn et al., 1993). This high degree of linkage conservation and the fact that probes can be used between species (Tanksley et al., 1992) will doubtlessly accelerate integrative mapping strategies.

We are engaged in the transfer of molecular approaches of genome analysis to the cytogenetic dimension. During the past five years light microscopic cytological mapping strategies based on marker techniques have been developed, for instance, genomic in situ hybridization (GISH) to detect chromatin introgression from alien wild species into cultivars of commercial interest (Anamthawat-Jónsson et al., 1990; Mukai and Gill, 1991; Schwarzacher et al., 1992) or the mapping of polymorphic C-bands which allowed the study of recombination events along the chromosome (Curtis and Lukaszewski, 1991; Lukaszewski, 1992). In spite of this progress, the dimensions between the 'DNA thread' (macromolecule) on the one hand and the cytological range on the other, and their consequences for the processes of maintenance, evolution and expression of genetic information have largely escaped analysis. Their importance is illustrated by the enormous differences seen in the expression of a discrete transgene which are generally attributed to (ill-defined) 'position effects'. The recent development of the field emission scanning electron microscope in combination with non-radioactive probe hybridization (Lichter et al., 1990) offers unique perspectives and promises to bridge this gap in gene and genome research (Wanner et al., 1990, 1991; Martin et al., 1994, 1995).

SYNTENY ANALYSIS AND HIGH-RESOLUTION MAPPING

Comparative mapping studies

The polyploid character of wheat with its three compensating homoeologous genomes generally tolerates deletions or deficiencies induced by an *Aegilops cylindrica* chromosome (Endo, 1988; Endo and Gill, 1995). A monosomic *Aegilops cylindrica* chromosome addition line produces gametes with chromosome rearrangements, preferentially single terminal deletions, that are randomly distributed in the genome. This excellent material has been chosen to comparatively analyze and map Triticeae genomes (Gill et al., 1993; Kota et al., 1993; Werner et al., 1992). To date, more than 400 deletion lines, disomic for single terminal deletions, have been selected with C- or N-banding and light microscopic chromosome measurements (Endo and Gill, 1995). Group 7 chromosomes carried the most deletions among all homoeologous groups and were therefore chosen for detailed analysis (Hohmann et al., 1994; Werner et al., 1992). Sixty-three group 7 deletion lines were studied for the presence or absence of 111 RFLP and RAPD markers. Fifty-three regions (88.3%) out of 60 chromosome regions of chromosomes 7A, 7B or 7D were tagged with molecular markers (Hohmann et al., 1995a).

The close relationship between Gramineae species and the availability of a large number of RFLP markers should enable us to construct a high resolution physical consensus map of wheat (Hohmann et al., 1994, 1995a), and also to integrate this map with genetic linkage maps from other Gramineae species, for instance, barley. In fact, the integration of wheat, barley

and oat RFLP markers has proven the collinearity between the wheat A-, B- and D-genomes, the H-genome of barley (Hohmann et al., 1995a) and the E-genome of *Agropyron* (U. Hohmann et al., unpublished). Landmark loci, such as cDNAs and single- or low-copy genomic DNAs that correspond to highly conserved coding regions, are convenient tools to align different maps and to allocate orthologous loci among the Triticeae genomes. However, we have recently shown that 9 of 63 analyzed group 7 deletion lines possess multiple chromosome rearrangements (Hohmann et al., 1995b). The absence of proximal markers or the presence of distal markers relative to the deletion breakpoints in the short and long arms indicated the existence of additional interstitial deletions in those lines, which have thus been eliminated from the construction of the consensus map as they may change or obfuscate the linear order of markers.

The high-density gene mapping of defined chromosome regions will provide information for transfer strategies of orthologous loci of disease resistance and quantitative traits. It may also serve as an anchor for integrating large genome segments originating from megabase technologies like the YAC and BAC systems, and be of significance in understanding genome evolution among the Triticeae by analyzing structural rearrangements in chromosomes, recombination hot spots, suppression of recombination, gene distribution, gene duplication and elimination events. The identification of non-collinear markers, in turn, will lead to the exploration of duplicated loci, which often complicate linkage mapping studies.

High-density physical maps

Homologous RFLP and other markers that had been genetically mapped in barley were used to compare both the genetic and physical order of markers between barley and wheat. The physical maps of wheat are collinear to linkage maps of barley (Hohmann et al., 1995a). The order of 100 (85%) of the 118 loci, 53 on the short arm and 65 on the long arm that were physically mapped in wheat was identical with the linkage map of barley. The synteny of markers in the long arm is fully conserved. In the short arm, the order of four proximal markers (*Xpsr150/ Xmwg705/ Xmwg808/ Xabg466*) and that of the marker pair *Xabc151* and *Xmwg89* appears to be converse. The four proximal markers in the I/F linkage map and the pair of markers in the S/M linkage map are tightly linked. Of 39 RFLP markers selected with: (i) an even genetic distribution along the chromosome; and (ii) a clustered arrangement around the centromere, 18 DNA fragments detected no polymorphism between the individual genomes of wheat, eight of which were expected to identify loci in the centromere region. Obviously, centromere segments of barley are highly conserved between both species which may impede the progress in the comparative mapping of centromere regions in Triticeae material. In comparative maps the density of common proximal markers is reduced. We estimated the distribution of recombination along barley chromosome 7H, which is one of the genetically best mapped chromosomes in the Triticeae (Graner et al., 1991). For this, each arm of the physical consensus map of wheat was divided into three (proximal, interstitial and distal) regions of approximately equal sizes. The marker density in the distal third of the short and long arms of barley chromosome 7H was two times higher than in the proximal region. The recombination frequencies in the distal region appear to be 8- to 15-fold enhanced compared to the respective proximal third of each arm. The relatively low recombination frequency in proximal and the increased one in distal chromosome regions, illustrated by Fig. 1, may have an impact on gene transfer across species.

The wheat genome consists of 16,000 million base pairs per haploid genome (Arumuganathan and Earle, 1991) and the entire length of the haploid set of chromosomes is 235.4 µm (Gill et al., 1991). The length of chromosome 7A is 11.3 µm (Endo and Gill, 1984). To date, 29 different chromosome segments have been distinguished on chromosome 7A (Endo and Gill, 1995) implying the average size is 0.39 µm per region. This corresponds to twice the resolution of the light microscope. With 95 RFLP loci mapped on chromosome 7A the average number of loci per region is 3.3. The average marker distance is 0.12 µm or 8.1 Mb. The highest marker density detected resides in the long arm at region FL 0.86-0.87 with 6 markers. In that region markers have been mapped at a 0.01 µm distance or every 800 kb. Any gene of interest in such a region would thus be accessible to map-based cloning strategies.

Microdissection and subchromosomal libraries

The construction of high-resolution maps from discrete genome regions can greatly facilitate genetic finger-printing, gene isolation and QTL studies. Therefore, we have worked out a refined microdissection procedure and applied it to monosomic as well as to telosomic chromosomes in telotrisomic or ditelotetrasomic barley and telosomic wheat/barley addition lines (Schondelmaier et al., 1993). The conventional cloning of subpicogram quantities of DNA either by flow sorting (Conia et al., 1987; Wang et al., 1992), or microdissection of metaphase chromosomes via glass needles (Lüdecke et al., 1989) or laser beams (Hadano et al., 1991) has been generally hampered by the availability of sufficient chromosome numbers, the purity of chromosomes, and a low cloning efficiency. Microdissection and construction of chromosomal and sub-chromosomal libraries has been significantly improved by the use of: (i) mitotic metaphase spreads of synchronized meristematic root tip tissue from which metaphase spreads of appropriate quality can be readily obtained; (ii) fixation times of less than three minutes that are possible with a novel drop-spread technique; (iii) a specific vector (Lüdecke et al., 1989) in combination with nanolitre-scale PCR that allows us to clone the DNA of 10 or

less chromosomes or chromosome segments; (iv) mineral oil rather than a moist chamber to prevent evaporation from the collection drop; and (v) an instrumental set-up which allows us to select suitable metaphases, to store the coordinates of their positions, and to address sequentially individual metaphases or the collection drop automatically (Schondelmaier et al., 1993). The latter facility is particularly advantageous when dissection attempts the isolation of several fragments per chromosome arm, or chromosomes per cell. The outlined approach is capable of directly cloning the DNA of individual chromosomes or chromosome segments which are well discernible by their morphology, for instance, telosomic chromosomes in mitotic metaphase spreads. Such subchromosomal libraries are, in principal, available for the set of

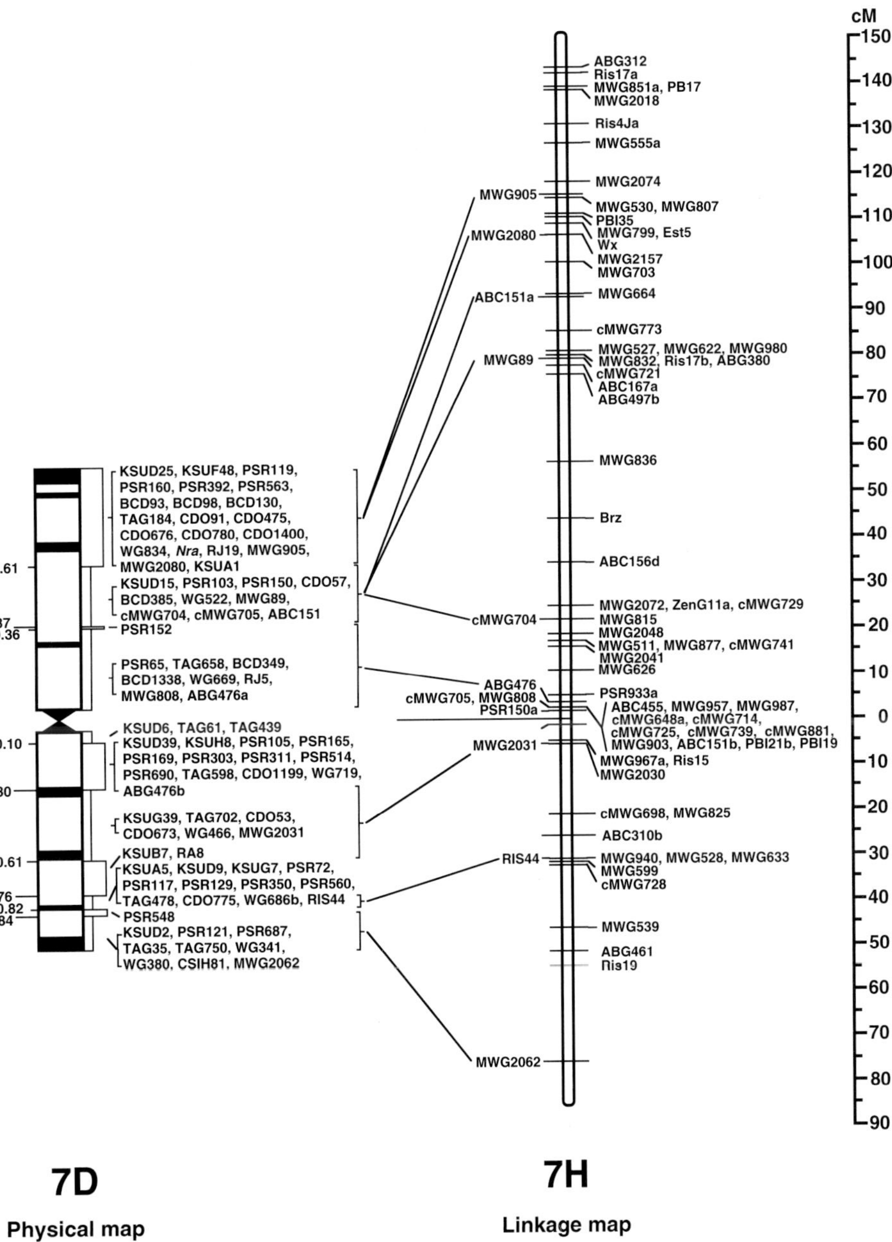

Fig. 1. Comparison of the physical map of C-banded wheat chromosome 7D and the linkage map from barley chromosome 7H. On the physical map the fractional lengths (FLs) are shown on the left and the marker positions on the right. The linkage map is based on the linkage I/F (Igri/Franka) map.

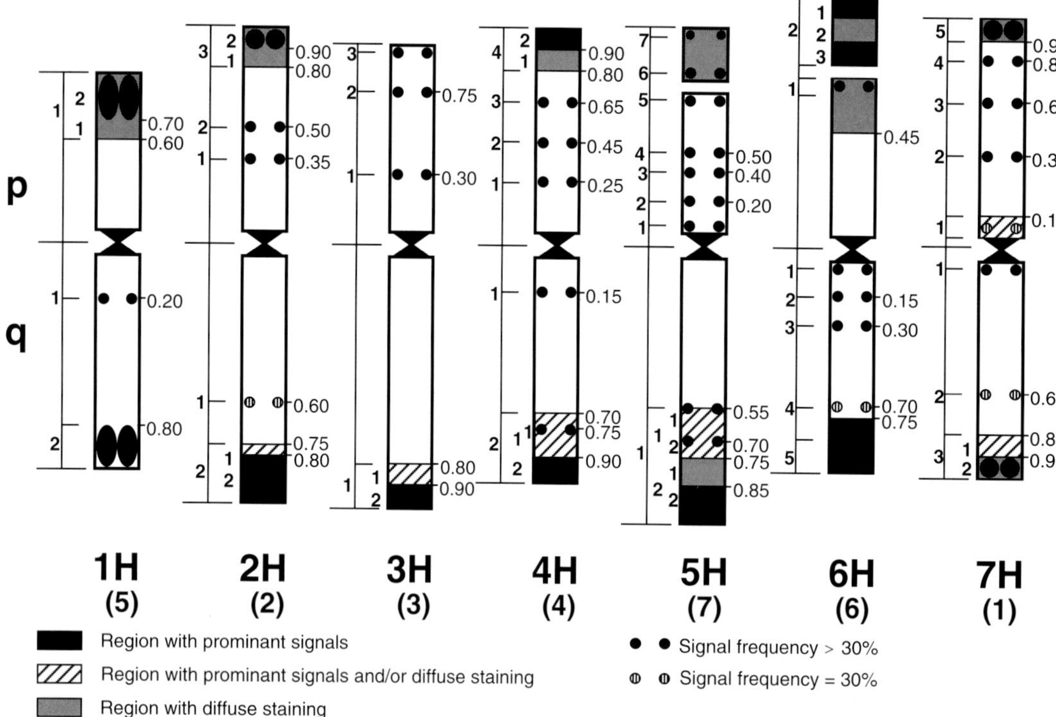

Fig. 2. Idiogram of labelled *Hordeum vulgare* chromosomes obtained with probe pHvMWG2315. The number of labelled regions is indicated on the left and FL (fractional length) positions on the right of each chromosome.

telosomes covering the entire barley genome. The representative plasmid library of chromosome arm 1HS with 44,000 recombinant clones has an average insert size of 250 bp. The frequency of repetitive clones was 60%, and 3.6% of the sequences isolated were polymorphic.

Light microscopic in situ techniques

Comparative mapping of low copy sequences

The efficient preparation of plant chromosomes for light microscopy (Busch et al., 1994) and high resolution scanning electron microscopy (Martin et al., 1994) employing an improved drop technique for plant protoplasts and the sensitivity enhancement of fluorescence in situ hybridization (FISH) can accelerate the comparative topographical mapping of plant genomes (Busch et al., 1995a,b).

With FISH the *Sec-1* locus (ω-secalin) of rye has been mapped on the satellite of 1RS. The genomic probe pSec2B was allocated in the middle of the satellite, distal to its heterochromatic and proximal to its euchromatic area (Busch et al., 1995a). Double hybridization with the 26 S rDNA probe CHS50 (W. Busch et al., unpublished data) from barley that covers the nucleolus organisator region (NOR) demonstrated that the *Sec-1* locus is clearly discernible from the NOR in the satellite. The distance between *Sec-1* and the NOR is approximately five times longer than that between *Sec-1* and the heterochromatic region. Heterologously, pSec2B hybridized to chromosome arm 1HS in barley near the telomere region that harbors the syntenic B-hordein locus (Lehfer et al., 1993). Chromosome 1H is sub-metacentric and the smallest barley chromosome.

Karyotyping by comparative mapping of repetitive sequences

To unambiguously identify every individual barley chromosome in a single hybridization step (Fig. 2) an ISH was estab-

lished with tandemly amplified repetitive sequence elements from microdissected barley chromosomes (Fig. 2; Busch et al., 1995b). The tandem repeat unit of probe pHvMWG2315 is approximately 330 bp long. It shares between 75% and 85% sequence similarity with other known repetitive sequences in the Triticeae. Together with pAs1 from *Triticum tauschii* (Rayburn and Gill, 1986), dpTa1 from *Triticum aestivum* (Metzlaff et al., 1986; Vershinin et al., 1994) and pHcKB6 from *Hordeum chilense* (Anamthawat-Jónson and Heslop-Harrison, 1993), the pHvMWG2315 repeat of *Hordeum vulgare* (Busch et al., 1995b) is a representative of a tandem repeat family in the Triticeae that is highly amplified in the genomes of *T. tauschii*, *H. vulgare* and *Agropyron elongatum*. These sequences have been used as a diagnostic tool to identify genomes either in ISH karyotyping of single Gramineae genomes with pAs1 (Rayburn and Gill, 1986; Mukai et al., 1993; Cabrera et al., 1995) and with pHvMWG2315 (Busch et al., 1995b) or in comparative RFLP studies with the dpTa1 probe (Vershinin et al., 1994). Sequence analysis of DNA elements obtained from PCR with genomic DNA from *T. tauschii* and *H. vulgare* has uncovered sub-regions with significant 'species-specific' sequence differences that can be used for homologous and heterologous ISH at different stringency conditions. For example, in heterologous hybridization with the *T. tauschii* pAs1 sequence the identification of all barley chromosomes at low stringency (70%) was possible. No specific signals were visible at high stringency (80%). The pHvMWG2315 sequence from *H. vulgare* which displays karyotyping capability at low stringency on D genome chromosomes in wheat as well, hybridizes to the three major sites at high stringency (80%). It is important to note that this approach is also useful for simultaneous double or triple hybridizations in an attempt to directly assign to and localize new sequences on specific chromosomes or chromosome segments in a single step (Busch et al., 1995b).

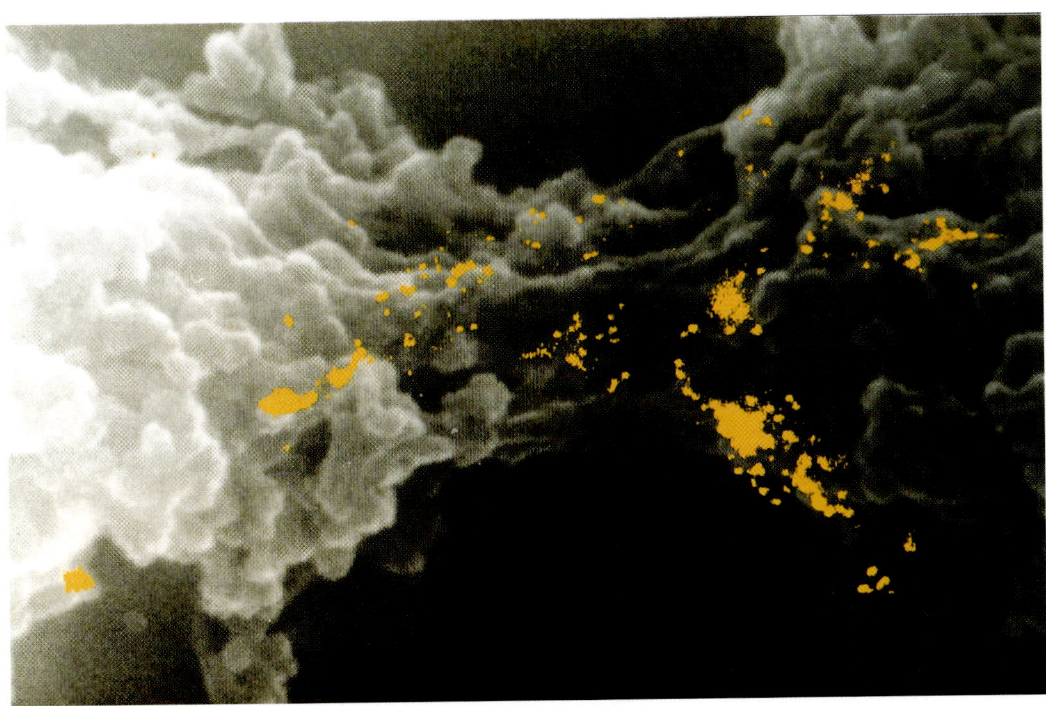

Fig. 3. Scanning electron micrograph of a secondary constriction of barley chromosome 5H labelled with an rDNA sequence (gold particles).

Physical mapping of sequences by high resolution electron microscopy

Fine allocation of rDNA sequences

We have recently succeeded in applying the ISH techniques at the ultrastructural level *without* significant chromosome structural changes (Martin et al., 1995). In situ hybridization and signal detection by high resolution field emission scanning electron microscopy allowed the allocation of colloidal gold particles on chromosome structures resembling the 30 nm fibre at the secondary constriction and compact adjacent regions of barley chromosomes (Fig. 3).

This work was supported by grants from the Bundesministerium für Forschung und Technologie (grant 0318990F), the Deutsche Forschungsgemeinschaft and the Fonds der Chemischen Industrie.

REFERENCES

Ahn, S. N. and Tanksley, S. D. (1993). Comparative linkage maps of the rice and maize genomes. *Proc. Nat. Acad. Sci. USA* **90**, 7980-7984.

Ahn, S. N., Anderson, J. A., Sorrells, M. E. and Tanksley, S. D. (1993). Homoeologous relationships of rice, wheat and maize chromosomes. *Mol. Gen. Genet.* **241**, 483-490.

Anamthawat-Jónsson, K., Schwarzacher, T., Leitch, A. R., Bennett, M. D. and Heslop-Harisson, J. S. (1990). Discrimination between closely related *Triticeae* species using genomic DNA as a probe. *Theor. Appl. Genet.* **79**, 721-728.

Anamthawat-Jónsson, K. and Heslop-Harisson, J. S. (1993). Isolation and characterization of genome-specific DNA sequences in Triticeae species. *Mol. Gen. Genet.* **240**, 151-158.

Arumuganathan, K. and Earle, E. D. (1991). Nuclear DNA content of some important plant species. *Plant. Mol. Rep.* **9**, 208-218.

Berhan, A. M., Hulbert, S. C., Butler, L. G. and Bennetzen, J. L. (1993). Structure and evolution of the genomes of *Sorghum bicolor* and *Zea mays*. *Theor. Appl. Genet.* **86**, 598-604.

Bonierbale, M., Plaisted, R. L. and Tanksley, S. D. (1988). RFLP maps of potato and tomato based on a common set of clones reveal modes of chromosomal evolution. *Genetics* **120**, 1095-1103.

Busch, W., Martin, R. and Herrmann, R. G. (1994). Sensitivity enhancement of fluorescence in situ hybridization on plant chromosomes. *Chrom. Res.* **2**, 15-20.

Busch, W., Herrmann, R. G. and Martin, R. (1995a). Refined physical mapping of the *Sec-1* locus on the satellite of chromosome 1R of rye. *Genome* **38**, 889-893.

Busch, W., Martin, R., Herrmann, R. G. and Hohmann, U. (1995b). Repeated DNA sequences isolated by microdissection. I. Karyotyping of barley (*Hordeum vulgare* L.). *Genome* **38**, 1082-1090.

Cabrera, A., Friebe, B., Jiang, J. and Gill, B. S. (1995). Characterization of *Hordeum chilense* by C-banding and in situ hybridization using highly repeated DNA probes. *Genome* **38**, 435-442.

Conia, J., Bergounioux, C., Perennes, C., Muller, P., Brown, S. and Gadal, P. (1987). Flow cytometric analysis and sorting of plant chromosomes from Petunia hybrida protoplasts. *Cytometry* **8**, 500-508.

Curtis, C. A. and Lukaszewski, A. J. (1991). Genetic linkage between C-bands and storage protein genes in chromosome 1B of tetraploid wheat. *Theor. Appl. Genet.* **81**, 245-252.

Endo, T. R. (1988). Induction of chromosomal structural changes by a chromosome of *Aegilops cylindrica* L. in common wheat. *J. Hered.* **79**, 366-370.

Endo, T. R. and Gill, B. S. (1984). Somatic karyotype, heterochromatin distribution and nature of chromosome differentiation in common wheat, *Triticum aestivum* L. em Thell. *Chromosoma* **89**, 361-369.

Endo, T. R. and Gill, B. S. (1995). Production of deletion stocks in common wheat. In (ed. Z. S. Li and Xin, Z. Y.) Proc. 8th Int. Wheat Genet. Symp. Bejing, China, **1**, 211-216.

Gill, B. S., Friebe, B. and Endo, T. R. (1991). Standard karyotype and nomenclature system for description of chromosome bands and structural aberrations in wheat (*Triticum aestivum*). *Genome* **34**, 830-839.

Gill, K. S., Gill, B. S. and Endo, T. R. (1993). A chromosome region-specific mapping strategy reveals gene-rich telomeric ends in wheat. *Chromosoma* **102**, 374-381.

Graner, A., Jahoor, A., Schondelmaier, J., Siedler, H., Pillen, K., Fischbeck, G., Wenzel, G. and Herrmann, R. G. (1991). Construction of an RFLP map of barley. *Theor. Appl. Genet.* **83**, 250-256.

Hadano, S., Watanabe, M., Yokoi, H., Kogi, M., Kondo, I., Tsuchiya, H., Kanazawa, I, Wakasa, K. and Ikeda, J.-E (1991). Laser microdissection and single unique primer PCR allow generation of regional chromosome DNA clones from a single human chromosome. *Genomics* **11**, 364-373.

Hohmann, U., Endo, T. R., Gill, K. S. and Gill, B. S. (1994). Comparison of genetic and physical maps of group 7 chromosomes from *Triticum aestivum* L. *Mol. Gen. Genet.* **245**, 644-653.

Hohmann, U., Graner, A., Endo, T. R., Gill, B. S. and Herrmann, R. G. (1995a). Comparison of wheat physical maps with barley linkage maps for group 7 chromosomes. *Theor. Appl. Genet.* **91**, 618-626.

Hohmann, U., Endo, T. R., Herrmann, R. G. and Gill, B. S. (1995b).

Characterization of deletions in common wheat induced by an *Aegilops cylindrica* chromosome: detection of multiple chromosome rearrangements. *Theor. Appl. Genet.* **91**, 611-617.

Hulbert, S. H., Richter, T. E., Axtell, J. D. and Bennetzen, J. L. (1990). Genetic mapping and characterization of *Sorghum* and related crops by means of maize DNA probes. *Proc. Nat. Acad. Sci. USA* **87**, 4251-4255.

Jordan, B. R. (1988). Megabase methods: a quantum jump in recombinant DNA techniques. *BioEssays* **8**, 140-145.

Kota, R. S., Gill, B. S. and Endo, T. R. (1993). A cytogenetically based physical map of chromosome 1B in common wheat. *Genome* **36**, 548-554.

Lehfer, H., Busch, W., Martin, R. and Herrmann, R. G. (1993). Localization of the B-hordein locus on barley chromosomes using fluorescence in situ hybridization. *Chromosoma* **102**, 428-432.

Lichter, P., Chang Tang, C. and Call, K. (1990). High-resolution mapping of human chromosome 11 by in situ hybridization with cosmid clones. *Science* **24**, 64-69.

Lüdecke, H. J., Senger, G., Claussen, U. and Horsthemke, H. J. (1989). Cloning defined regions of the human genome by microdissection of banded chromosomes and enzymatic amplification. *Nature* **338**, 348-350.

Lukaszewski, A. J. (1992). A comparison of physical distribution of recombination in chromosome 1R in diploid rye and in hexaploid Triticale. *Theor. Appl. Genet.* **83**, 1048-1053.

Martin, R., Busch, W., Herrmann, R. G. and Wanner, G. (1994). Efficient preparation of plant chromosomes for high resolution scanning electron microscopy. *Chrom. Res.* **2**, 411-415.

Martin, R., Busch, W., Herrmann, R. G. and Wanner, G. (1995). In situ hybridization and signal detection by high resolution scanning electron microscopy. *Kew Chromosome Conference 4* (ed. P. E. Brandham and M. D. Bennett), pp. 159-166. Royal Botanical Gardens, Kew, UK.

Metzlaff, M., Troebner, W., Baldauf, F., Schlegel, R. and Cullum, J. (1986). Wheat specific repetitive DNA sequences - construction and characterization of four different genomic clones. *Theor. Appl. Genet.* **72**, 207-210.

Mukai, Y., Nakahara, Y. and Yamamoto, M. (1993). Simultaneous discrimination of the three genomes in hexaploid wheat by multicolor fluorescence in situ hybridization using total genomic and highly repeated DNA probes. *Genome* **36**, 489-494.

Mukai, Y. and Gill, B. S. (1991). Detection of barley chromatin added to wheat by genomic in situ hybridization. *Genome* **34**, 448-452.

Osborn, T. C., Alexander, D. C. and Fobes, J. F. (1987). Identification of restriction fragment length polymorphisms linked to genes controlling soluble solids content in tomato fruit. *Theor. Appl. Genet.* **73**, 350-356.

Paterson, A. H., Lander, E. S., Hewitt, J. D., Peterson, S., Lincoln, S. E. and Tanksley, S. D. (1988). Resolution of quantitative traits into Mendelian factors by using a complete linkage map of restriction fragment polymorphisms. *Nature* **335**, 721-726.

Rayburn, A. L. and Gill, B. S. (1986). Isolation of a D-genome specific repeated DNA sequence from *Aegilops squarrosa. Plant Mol. Biol. Rep.* **4**, 102-109.

Saiki, R. K., Scharf, S., Falcona, F. Mullis, K. B., Horn, G. T., Erlich, H. A. and Arnheim, N. (1985). Enzymatic amplification of β-globulin genomic sequences and restriction site analysis for diagnosis of sickle cell anemia. *Science* **230**, 1350-1354.

Schondelmaier, J., Martin, R., Jahoor, A., Houben, A., Graner, A., Koop, H. U., Herrmann, R. G. and Jung, C. (1993). Microdissection and microcloning of the barley (*Hordeum vulgare* L.) chromosome 1HS. *Theor. Appl. Genet.* **86**, 629-636.

Schwarzacher, T., Anamthawat-Jónsson, K., Harrison, G. E., Islam, A. K. M. R., Jia, J. Z., King, I. P., Leitch, A. R., Miller, T. E., Reader, S. M., Rogers, W. J., Shi, M. and Heslop-Harrison, J. S. (1992). Genomic in situ hybridization to identify alien chromosomes and chromosome segments in wheat. *Theor. Appl. Genet.* **84**, 778-786.

Tanksley, S. D., Ganal, M. W., Price, J. P., De Vicente, M. C., Bonierbale, M. W., Broun, P., Fulton, T. M., Giovanonni, J. J., Grandillo, S., Martin, G. B., et al. (1992). High density molecular linkage maps of the tomato and potato genomes; biological interferences and practical approaches. *Genetics* **132**, 1141-1160.

Vershinin, A., Svitashev, S., Gummesson, P.-O., Salomon, B., Von Bothmer, R. and Bryngelsson, T. (1994). Characterization of a family of repeated DNA sequences in Triticeae. *Theor. Appl. Genet.* **89**, 217-225.

Wang, M. L., Leitch, A. R., Schwarzacher, T., Heslop-Harisson, J. S. (1992). Construction of a chromosome-enriched HpaII library from flow-sorted wheat chromosomes. *Nucl. Acids Res.* **20**, 1897-1901.

Wanner, G., Formanek, H. and Herrmann, R. G. (1990). Ultrastructure of plant chromosomes by high-resolution scanning electron microscopy. *Plant Mol. Biol Rep.* **8**, 224-236.

Wanner, G., Formanek, H., Martin, R. and Herrmann, R. G. (1991). High-resolution scanning electron microscopy of plant chromosomes. *Chromosoma* **100**, 103-109.

Werner, J. E., Endo, T. R., Gill, B. S. (1992). Towards a cytogenetically based physical map of the wheat genome. *Proc. Nat. Acad. Sci. USA* **89**, 11307-11311.

Whitkus, R., Doebley, J. and Lee, M. (1992). Comparative genome mapping of *Sorghum* and maize. *Genetics* **132**, 1119-1130.

Comparative genetic and QTL mapping in sorghum and maize

Michael Lee

Iowa State University, USA

SUMMARY

DNA markers and genetic maps will be important tools for direct investigations of several facets of crop improvement and will provide vital links between plant breeding and basic plant biology. The markers and maps will become more important for increased crop production because plant genetics will be required to extend or replace extant management practices such as chemical fertilizers, pesticides, and irrigation (Lee, 1995). Despite the importance of the sorghum crop, comprehensive genetic characterization has been limited. Therefore, the primary goal of this research program was to develop basic genetic tools to facilitate research in the genetics and breeding of sorghum. The first phase of this project consisted of constructing a genetic map based on restriction fragment length polymorphisms (RFLPs). The ISU sorghum map was created through linkage analysis of 78 F_2 plants of an intraspecific cross between inbred CK60 and accession PI229828 (Pereira et al., 1994). The map consists of 201 loci distributed among 10 linkage groups covering 1,299 cM. Comparison of sorghum and maize RFLP maps on the basis of common sets of DNA probes revealed a high degree of conservation as reflected by homology, copy number, and collinearity. Examples of conserved and rearranged locus orders were observed. The same sorghum population was used to map genetic factors (mutants and QTL) for several traits including vegetative and reproductive morphology, maturity, insect, and disease resistance. This presentation will emphasize analysis of genetic factors affecting plant height, an important character for sorghum adaptation in temperate latitudes for grain production. Four QTL for plant height were identified in a sample of 152 F_2 plants (Pereira and Lee, 1995) whereas 6 QTL were detected among their F_3 progeny. These observations and assessments of other traits at 4 QTL common to F_2 plants and their F_3 progeny indicate some of these regions correspond to loci (*dw*) previously identified on the basis of alleles with highly qualitative effects. Four of the six sorghum plant height QTL seem to be orthologous to plant height QTL in maize. Other possible instances of orthologous QTL included regions for maturity and tillering. These observations suggest that the conservation of the maize and sorghum genomes encompasses sequence homology, collinearity, and function. The genetic information and technology developed on the basis of DNA markers could be used in several facets of breeding, genetics, and other basic biological investigations. In addition, DNA markers have been used to survey large collections of elite sorghum germ plasm to determine the degree of genetic relationships and genetic diversity (Ahnert et al., 1996). RFLP data seem to portray genetic relationships more accurately than the methods based exclusively on the coancestry coefficient. This information provides the basis for more accurate perceptions of genetic relationships and diversity.

Key words: Sorghum, Maize, RFLP, QTL, Genetics

INTRODUCTION

Knowledge of a crop's genetic architecture will become more important for increased crop production because plant genetics will be required to extend or replace extant management practices such as chemical fertilizers, pesticides, and irrigation. Such knowledge will include more detailed descriptions of genome organization, the crop's gene pools, and genes and pathways controlling important phenotypes. In many instances, DNA markers and genetic maps will be important tools for direct investigations of these areas and will provide vital links between plant breeding and basic plant biology (Lee, 1995).

Many of the limitations of plant breeding methods have been rooted in the status of the technical infrastructure for conducting genetic analyses. Breeders and geneticists of all crops have lacked an informative and integrated genetic context to aid interpretation and conciliation of perspectives provided by seemingly different approaches to genetic improvement. The result has been a situation resembling the Tower of Babel with breeders, geneticists, cytogeneticists, taxonomists, molecular biologists, plant pathologists,and other factions contributing to the confusion. A key component of the infrastructure and context of future plant breeding programs will be genetic maps. The maps, when fully integrated, will have several roles: (1) to provide a focal point and hub for data derived from the perspectives of myriad disciplines for each crop; (2) to constitute a vital two-way avenue connecting plant breeding and basic plant biology; (3) to contribute essential information for positional cloning of genes; (4) to facilitate a considerable and directed expansion of a crop's gene pool through comparative mapping of related and unrelated taxa; (5) to accelerate identification and incorporation of useful genes into cultivars; and

(6) to contribute important clues toward understanding the biological basis of complex traits and phenomena important to crop improvement. The significance of these and other roles and their implementation will vary with the repertoire of genetic technologies available to the crop, breeding methods and goals, and the nature of the crop's nuclear genome. However, the foundation provided by the maps will have a positive impact on the genetic improvement of crop species in many instances.

DEVELOPMENT OF INTEGRATED MAPS

Historically, genetic maps have provided very few advantages to plant breeding programs and crop improvement even for species such as maize, wheat, rice, and tomato with relatively well-developed maps. The primary problems have been the types of markers used predominantly to create maps (macromutations and cytological markers), existence of poorly integrated maps each based on a different type of marker, the lack of informative markers in germ plasm used by breeders, and the polyploid nature of many crop genomes. The current and future generations of maps ameliorate these problems in significant ways.

One of the initial stages of recent map integration has involved cytological and DNA markers. When adequate cytogenetic stocks and manipulations have been available, their union with DNA markers has significantly enhanced our perception of genome architecture for crops such as tomato and potato (Tanksley et al., 1992), maize (Weber and Helentjaris, 1989), wheat (Werner et al., 1992), barley and rye (Devos et al., 1993a). These investigations have revealed patterns of genome duplication, recombination, and cytogenetic-genetic distances along chromosomes. This information is essential for efficient deployment of a wide spectrum of genetic technologies from targeted cloning of important genes through introgression of exotic germ plasm (Devos et al., 1993b).

Another phase of map integration has involved DNA markers and macromutations. Reports of genetic linkage between DNA markers and macromutations have increased at a seemingly exponential rate. To the extent such alleles have been used in breeding programs, these reports comprise a considerable array of expanded opportunities for using markers as indirect selection criteria. An area of potentially more pervasive significance for crop improvement has been integrative mapping of partly-sequenced cDNA clones in crops such as maize (e.g. Chao et al., 1994). Especially for macromutation-rich maps this activity will provide many opportunities for matching mutants, collected and characterized over several decades with recently isolated molecules from various taxa. Eventually, this process will provide a basis for determining the biophysical basis of genetic variation and phenotypic expression for many traits. To the extent function and DNA sequence have been conserved across plant taxa (Helentjaris, 1993), integrated maps and their markers may represent a very important plant genetic resource for crop breeding.

The third, and least complete phase of integrative mapping involves polygenes or quantitative trait loci (QTL; Tanksley, 1993). Despite the inherent ambiguities of the process, QTL mapping will provide vital information for basic and applied aspects of crop improvement. QTL mapping conducts a pangenomic assessment of gene location and action for potentially any phenotype (i.e. trait). Several aspects of QTL mapping make this approach especially powerful for adding important genes and regions to maps: (1) relatively comprehensive coverage of the genome provided by DNA markers; (2) choice of mapping parents is extensive provided adequate DNA polymorphism may be detected. Therefore, important genetic regions may be added to maps in a directed manner depending on the ability to assess genetic diversity and variation for the trait(s). Unlike mapping macromutations, this approach may provide a more sensitive survey of the genome because it does not require fortuitous observation and recovery of alleles with highly qualitative effects on the trait(s) of interest.

COMPARATIVE MAPPING

The use of common sets of DNA probes to detect and map homologous sequences across sexually isolated species has revealed a surprisingly high degree of conservation in terms of copy number and homology of low copy probes, linkage, and locus order. Recognition of the considerable conservation of these features within sets of plants such as rice, wheat, and maize (Ahn et al., 1993); sorghum and maize (Pereira et al., 1994); wheat, barley, and rye (Devos et al., 1993a); tomato, potato, and pepper (Tanksley et al., 1988, 1992); *Arabidopsis* and *Brassica* (Teutonico and Osborn, 1994) has inspired the suggestion of considering such groups as single genetic systems (Helentjaris, 1993; Bennetzen and Freeling, 1993).

This concept should have considerable merit and mutual advantages for breeders and geneticists. Often, the genome size of one member of the group is several fold smaller than other members. The smaller genome size should accelerate positional cloning of orthologous genes. Once the gene in the source species has been cloned and sequenced, this information may be used to quickly isolate the orthologous gene in the target species as demonstrated by the isolation of the gene for chalcone flavonone isomerase in maize using sequence information from *Petunia*, snapdragon, and bean (Grotewold and Peterson, 1994). Also, the repertoire and number of mapped and characterized genes may vary considerably between members of a group (e.g. tomato vs potato, maize vs sorghum). In these instances, map information from the 'gene rich' species may provide important clues about a map region's genetic content in the 'gene poor' species and vice versa. Comparisons of locus order and distribution of recombination events may also elucidate barriers and suggest strategies to incorporate germ plasm in wide crosses (Devos et al., 1993b). For plant breeding programs, this information represents an opportunity for a considerable and directed expansion and improved definition of a crop's gene pool.

Comparative mapping of DNA clones has provided the basis for parallel investigations of other genetic factors. For example, a region containing a locus that conditions the absence of ligules has been conserved among rice, wheat and maize (Ahn et al., 1993). Similar inspections of linkage data of other taxa should reveal many other examples such as the parallel linkage between genes for resistance to leaf rust (*Puccinia* spp.) and prolamines in oats, wheat, and maize (Rayapati et al., 1994). Recently, the pattern of conserved linkage and function has been extended to include QTL. The initial report of orthologous QTL noted that the RFLP loci with the greatest effects on seed weight in mung bean and cowpea were detected by the same clones (Fatokun et al., 1992). In a similar manner, comparative mapping in maize and sorghum has revealed four putatively orthologous regions for plant height (Fig. 1; Pereira and Lee, 1995) and other traits. In sorghum, each region has a major effect on that trait and on a unique suite of other traits (e.g. tillering, panicle dimensions, leaf length and width) much like some of the *dw* loci in sorghum. Interestingly, plant height mutants at maize genetic loci in related regions have pleiotropic effects on some of the same combinations of traits as the sorghum QTL and the candidate *dw* loci.

population for several reasons: (1) adequate DNA polymorphism; (2) parental difference for resistance to an important insect pest, the aphid *Schizapus graminum* (greenbug, race E); (3) parental difference for resistance to the fungal pathogen, *Peronsclerospora sorghi*, the causal agent of downy mildew; (4) parental divergence for numerous morphological and developmental traits (e.g. plant height, flowering, and tillering); (5) no known chromosomal polymorphisms; and (6) presumably, rates and patterns of recombination more representative of other intraspecific crosses. Selection of this population and use of maize DNA probes to detect RFLPs permitted rather efficient collection of extensive genetic information for sorghum and the means of relating the information to that of maize and other grasses.

DEVELOPMENT OF A GENETIC MAP FOR SORGHUM AND COMPARATIVE MAPPING WITH MAIZE

In terms of cultivated area, sorghum (*Sorghum bicolor* L. Moench) is the world's fifth leading cereal (Dogget, 1988). Despite the importance of the sorghum crop, comprehensive genetic characterization has been limited. Therefore, the primary goal of this research program was to develop basic genetic tools to facilitate research in the genetics and breeding of sorghum. The specific objectives were to: (1) develop a complete genetic linkage map for sorghum based on DNA markers; (2) identify the genetic locations of genes of interest to sorghum breeding programs; (3) integrate the sorghum genetic map(s) with maps of allied grass species; and (4) assess the genetic diversity and relationships of sorghum germ plasm.

In sorghum, several maps have been developed for various objectives by different research groups. These maps are being compared and integrated on the basis of probes exchanged among various laboratories (G. Hart, personal communication). The RFLP linkage map produced at ISU was created through linkage analysis of 78 F_2 plants of an intraspecific cross between inbred CK60 and accession PI229828 (Pereira et al., 1994). These parents were selected to produce the mapping

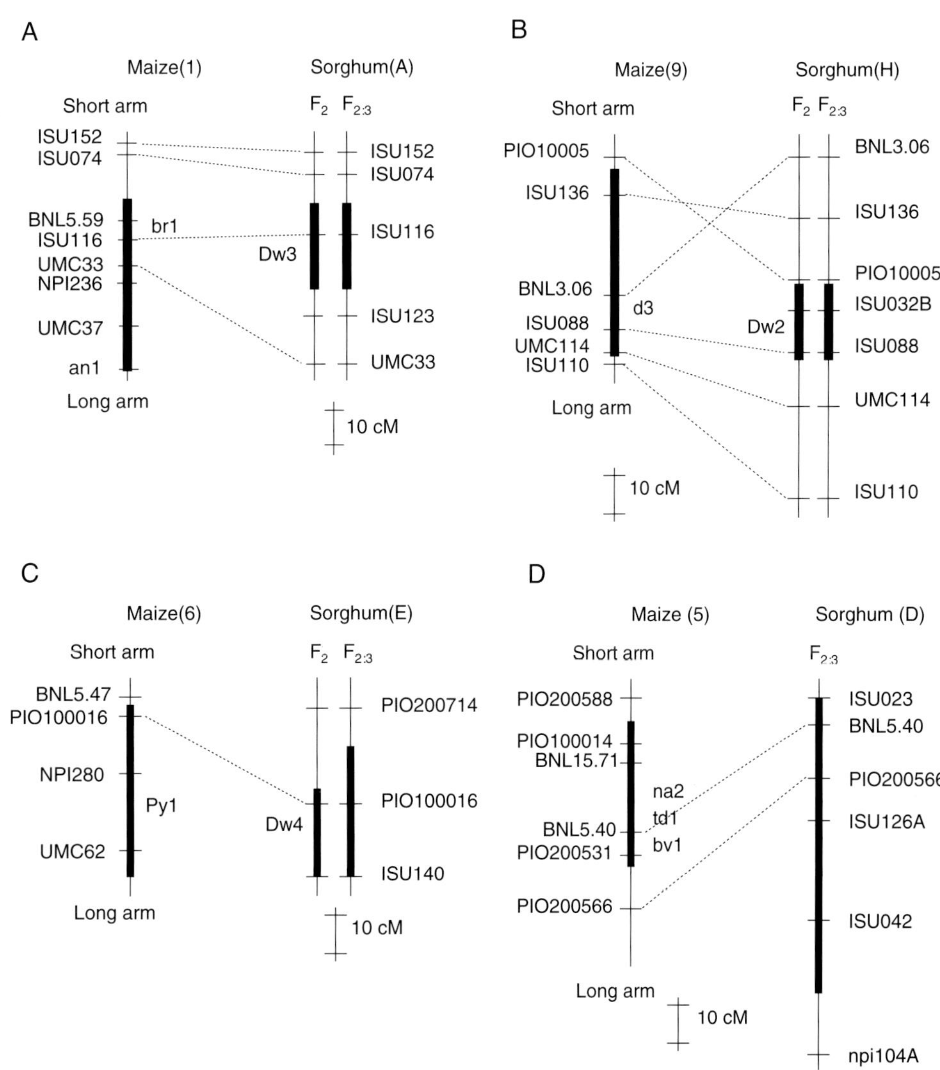

Fig. 1. Comparison of genomic regions of sorghum and maize having common RFLP loci (connected by dotted lines) linked to QTL for plant height. Numbers and letters in parenthesis represent the maize chromosome number and sorghum linkage group, respectively. The shaded areas are the confidence intervals (1.0 log unit as indicated by mapmaker-QTL program) for plant height QTL. In A,B and D the maize QTL were located by Beavis et al. (1991) and in C by Veldboom et al. (1994). *br1, an1, d3, Py1, na2, bv1* and *td1* represent maize loci with qualtitative mutants for plant height approximately located at that genomic region (Maize Newsletter, 1992). The positions of the *dw* loci are hypothetical.

The map consists of 201 loci distributed among 10 linkage groups covering 1,299 cM . Presumably the number of linkage groups correspond to the basic chromosome complement of sorghum (n=10). The RFLP loci were detected through hybridizations with probes of maize genomic (52), maize cDNA (124), and sorghum genomic (10) clones. Most probes detected a single RFLP locus (172) but there was evidence of genomic duplication as many probes (76) detected more than one band with a strong signal. However, in this population the additional bands were usually monomorphic. Segregation data at 95% of the loci fit expected ratios for an F_2 generation of a cross between two homozygous parents. Loci with deviant ratios were located predominantly to one region of linkage group B. All features of this initial map have been verified in a second and larger sample from this F_2 population used for mapping quantitative trait loci (QTL; Pereira and Lee, 1995).

The maize DNA clones used to construct maize RFLP maps, and subsequent mapping of new maize cDNA clones, permitted comparative linkage analysis. Maize and sorghum are both diploid (2n=2x=20) but the maize nuclear genome is 3-4 times larger than the sorghum genome. Comparison of sorghum and maize RFLP maps revealed a high degree of conservation as reflected by homology, copy number, and collinearity. Over 60% of the maize clones, genomic and cDNA, produce strong hybridization signals with sorghum. Many of the loci linked in maize (45 of 55 tested) were also linked in this sorghum population. Examples of conserved and rearranged locus orders were observed.

MAPPING MUTANTS AND QTL IN SORGHUM AND MAIZE

The same sorghum population was used to map genetic factors (mutants and QTL) for several traits including vegetative and reproductive morphology, maturity, insect, and disease resistance. This presentation will emphasize analysis of genetic factors affecting plant height, an important character for sorghum adaptation in temperate latitudes for grain production. Evaluations of the traits were conducted with 152 F_2 plants (Pereira and Lee, 1995) and their F_3 progeny (D. A. Ahnert, M. G. Pereira and M. Lee, unpublished). Analysis of the F_2 plants detected 4 unlinked QTL for plant height accounting for 63% of the variation. The QTL were located to linkage groups A, B, E, and H. Positive, additive genetic effects were estimated at 15-32 cm and alleles for increased stature were derived from the tall parent, PI229828. Tallness was dominant or overdominant at 3 QTL whereas short stature was dominant at the fourth on linkage group H. Epistasis was evident for one pair of QTL on linkage groups A and E. Analysis of F_3 progeny verified all features of the QTL detected in F_2 plants and detected 2 additional QTL for plant height in two other linkage groups, D and F. Analysis of other traits identified several QTL linked and unlinked to the plant height QTL. Two QTL for maturity (number of heat units to flowering) were identified as being linked to two plant height QTL. Regions initially identified as plant height QTL were associated with several other traits (from 1 to 10). Only the QTL on linkage group E was specific for plant height. Collectively, these data and assessments of other traits at the 4 QTL common to the F_2 plants and F_3 progeny indicate some of these regions correspond to loci

(dw) previously identified on the basis of alleles with highly qualitative effects. These observations and linkage relationships will be assessed with sets of near isogenic lines for each to the 4 dw loci.

COMPARATIVE QTL MAPPING

On the basis of integrated RFLP maps, the positions and effects of sorghum and maize plant height QTL were compared. Four of the six sorghum plant height QTL seem to be orthologous to plant height QTL in maize. The putative orthologous regions are (sorghum linkage group and maize chromosome) A and long arm of chromosome 1, D and chromosome 5, E and long arm of chromosome 6, H and chromosome 9. The regions of the maize plant height QTL also contain genetic loci defined by mutants with qualitative effects on stature such as *br1* and *an1* on chromosome 1, *na1* and *td1* on chromosome 5, *py1* on chromosome 6, and *d3* on chromosome 9. The effects of some of these maize mutants strongly resemble those of the sorghum plant height QTL and *dw* loci. At least 3 of those maize loci, *an1*, *br1*, and *d3* have been tagged with transposons or cloned by various laboratories. These sequences could be used to isolate the related gene from sorghum and further assess the degree and nature of conservation between these two genomes.

Comparative QTL analysis identified evidence for several other orthologous regions. For example, a region of linkage group A (*isu033-isu123*) was strongly associated with tillering and production of lateral branches. The LOD values were 2.8 and 8.7 in F_2 plants and F_3 families. As indicated by comparative mapping with RFLP loci and QTL for plant height, this region of the sorghum genome is most closely related to the long arm of chromosome 1 of maize. This region of the maize genome is the site of a genetic locus, *tb1*. The mutant phenotype at that locus is characterized by the production of many tillers and lateral branches in a manner strongly resembling the tillering QTL in sorghum. Other possible instances of orthologous QTL included regions for maturity. These observations suggest that the conservation of the maize and sorghum genomes encompasses sequence homology, collinearity, and function despite their divergence millions of years ago and subsequent evolution in different hemispheres with contrasting ecogeographical conditions.

SORGHUM BREEDING AND GENETICS WITH DNA MARKERS

The genetic information and technology developed on the basis of DNA markers could be used in several facets of breeding, genetics, and other basic biological investigations of sorghum. For example, sorghum breeders in temperate regions routinely transfer short stature and photoperiod insensitivity from adapted temperate inbred lines into exotic (tropical) germ plasm through several generations of backcrossing. The goal of the backcross breeding is to maximize the recovery of the tropical germplasm with a growth habit adapted to temperate latitudes and mechanized harvesting. Once the traits necessary for temperate adaptation have been adequately recovered, the converted tropical germ plasm may be evaluated in temperate regions for traits. Similar programs are conducted for many crops (e.g. rice, maize, and wheat), traits (e.g. resistance to diseases and insects, grain quality, male sterility), and for diversifying and enhancing gene pools (e.g. adapting temperate germplasm to tropical environments and vice versa).

The US Sorghum Conversion Program (Duncan et al., 1991) has used a backcrossing scheme to introduce genes for short stature and photoperiod insensitivity from temperate, adapted parent to the tropical, unadapted parent. According to that scheme, adapted segregants are selected on the basis of stature and ability to flower at the temperate latitude. Self-pollinated seed from such segregants is selected and used to backcross to the recurrent, tropical parent in a tropical environment. That backcross generation is self-pollinated in the tropical location and selfed seed is sown in the temperate location to identify segregants of appropriate stature and flowering response. Typically, 4 or more cycles of backcrossing, selfing, and selection are used to convert a tropical parent.

In this instance, DNA markers could be used to eliminate generations of backcrossing and to ensure that the tropical germplasm was indeed transferred. For example, DNA finger-prints of adapted segregants could be determined and used as a selection criteria prior to backcrossing. This should be especially effective for expediting recovery of chromosome regions from the tropical parent unlinked to the genes needed for adaptation to the temperate latitudes. For example, Pereira and Lee (1995) identified short-statured, adapted F_2 plants with RFLP genotypes closely resembling (60-75% similarity over 111 loci) the tall or short parent. By selecting the short plants with the greatest degree of resemblance to the tall parent, the backcross conversion could be completed in fewer generations. In addition, DNA marker loci could be used to minimize linkage drag from the adapted, donor parent in the vicinity of the genes affecting adaptation. Through such a scheme, the conservation and conversion of the tropical germ plasm would be optimized. Also, breeders would have the greatest opportunity to identify new and valuable genes from exotic germ plasm.

ASSESSING GENETIC DIVERSITY IN SORGHUM BREEDING PROGRAMS

One consequence of modern agricultural practices, which generally emphasizes maximum productivity with acceptable quality and uniformity, has been a reduction of genetic diversity of the primary gene pool under cultivation with similar fates for the secondary and tertiary gene pools of most major crops. That trend may be exacerbated in crops such as sorghum in which F1 hybrid seed is produced using cytoplasmic male sterility. Even though the extent of the reduction may be largely unquantifiable, it is generally assumed that valuable and irreplaceable genes have been lost or ignored, plant genetic resources have been shrinking at accelerated rates, and crop-based agriculture has become more vulnerable to the vagaries of climate and associated biotic and abiotic stresses. Undoubtedly there is considerable merit, validity, and controversy associated with each point. Facts and anecdotes aside, the consequences of a narrow genetic base of major crops have been experienced sporadically throughout recent history often with significant human and economic costs. Therefore, an awareness of genetic diversity and management of crop genetic resources has been an important component of plant improvement programs.

The prospects of utilizing DNA marker technology for managing germ plasm collections has been the subject of a recent and comprehensive review (Bretting and Widrlechner, 1995). Germ plasm management is a multifaceted endeavor involving acquisition, maintenance, and characterization such that the plant genetic resources are conserved and utilized for crop improvement. In the long term, conservation of collections probably deserves our greatest attention as the number of accessions and difficulties of preserving in situ reserves have increased for most crops while the risks to be managed have remained ambiguous, unpredictable, and very serious. Maintenance is likely to become more difficult as the financial costs of managing collections, especially of large, long-lived, perennial species, increases. However, the primary concerns of plant breeding programs involve issues of greater significance in the near term. Those of most immediate interest to plant breeding programs involve knowledge of the current genetic content of the collections – acquisition and the distribution of genetic diversity among accessions, relationships of collections (new and old) to elite germ plasm, and characterization of their potential genetic merit.

How do plant breeders assess genetic diversity and relationships among elite germ plasm? Many of the methods used by germ-plasm managers have been used by plant breeders (e.g. morphology and ecogeographical data). In addition, plant breeders often have access to pedigree information, performance records (e.g. combining ability, progeny evaluation, selection and breeding history), and inferences gleaned from various mating designs. The strength of some of the methods is that they are often based on direct assessments of what the breeder needs to know about the germ plasm. Such methods will be extremely difficult to improve. However, some methods and concepts have relied on weak genetic foundations, if any. For some plant breeding practices, that may constitute a weakness which reduces their efficiency. At least some of these deficiencies may be satisfied in part by DNA markers.

One of the most pervasive measures of genetic relationships in elite crop germ plasm has been Malecot's coefficient of coancestry (f), which provides an estimate of the degree of genetic similarity between two individuals (Melchinger, 1993). This measure estimates the probability that two randomly drawn, homologous genes (alleles) from each of two individuals are identical by descent. The measure has been based on Mendelian inheritance and probability and has been calculated under several assumptions: (1) absence of selection, mutation, migration, and drift; (2) regular diploid meiosis; and (3) no relationship (f=0) for individuals without verified common ancestors (Melchinger, 1993). Several common features of plant breeding programs have represented departures from these assumptions: (1) intense selection; (2) drift due to small sample sizes; (3) irregular nondiploid meiosis for some crops; and (4) unknown or incorrect pedigree records. Nevertheless, this method of estimating the degree of similarity will create information each generation and has been used by crop breeding programs.

DNA markers have been used to assess large collections of elite sorghum germ plasm to determine the degree of genetic relationships and genetic diversity (Fig. 2; Ahnert et al., 1996).

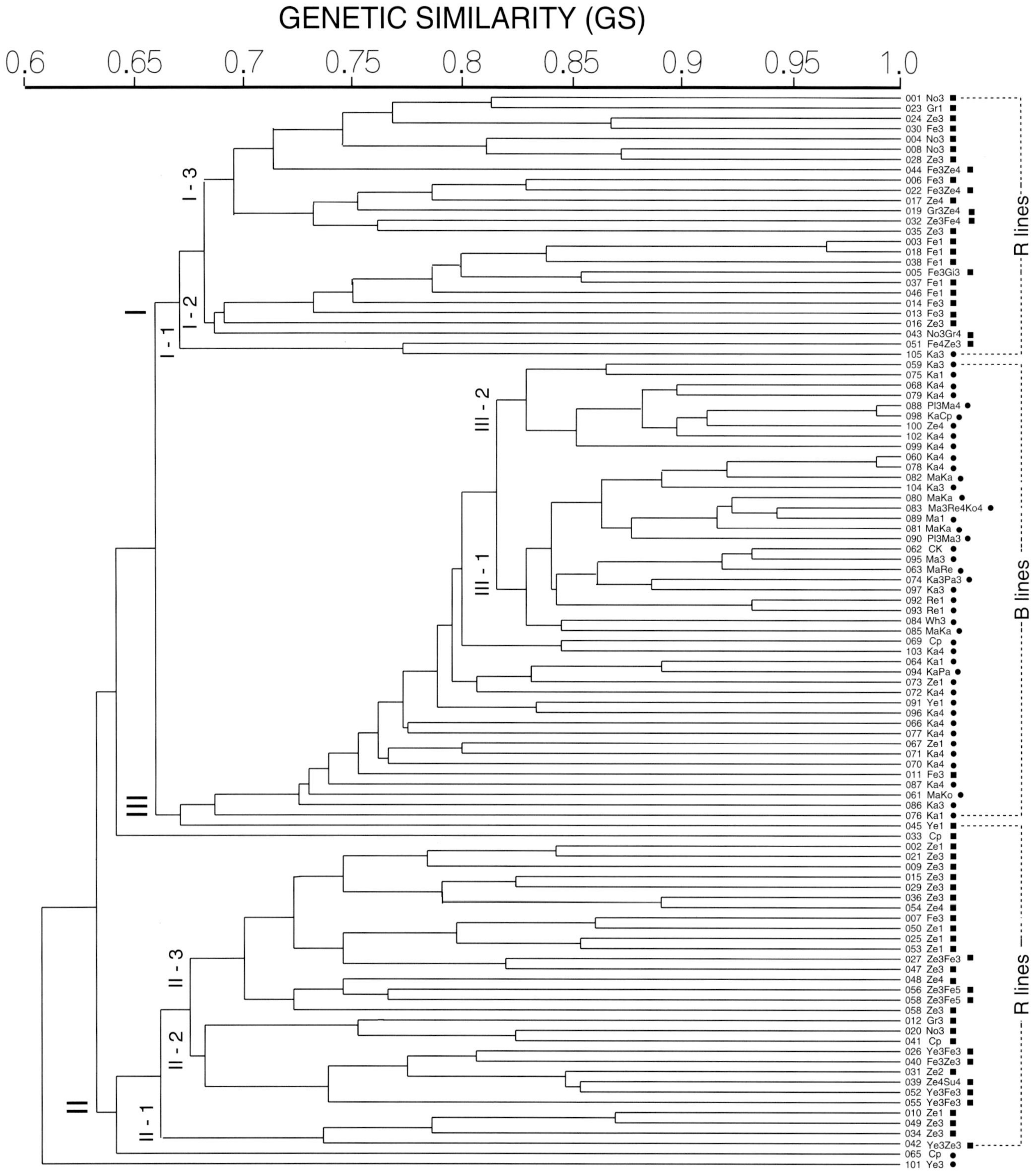

Fig. 2. Dendrogram of the 105 sorghum inbred lines revealed by UPGMA cluster analysis of RFLP-based genetic similarity (GS) estimates. ■, R lines; ●, B lines.

A set of 58 R lines, mostly from Kafir, and 47 B lines, mostly from Zera-zera or Feterita were surveyed with 104 DNA probes. The RFLP data clearly identify two different gene groups of inbred lines (pollen vs seed parents of F1 hybrids; R and B lines, respectively) and document the high degree of genetic similarity among members of certain gene pools (e.g. B lines). On average, 3.6 RFLP-band patterns per probe were observed for R lines whereas only 3 were detected for B lines. Estimates of genetic similarity based on RFLP fingerprints were 0.67 and 0.76 for R and B lines. Cluster analysis of RFLP data further divided R lines into 2 distinct groups representing derivatives from Feterita and Zera-zera.

Similar assessments may be made without DNA markers but the methods require extensive pedigree records and generations of breeding.. This requires generations of careful record keeping and is based on several assumptions mentioned previously. However, there was only a modest positive correlation between estimates of genetic similarity based on RFLP data and those based on the coancestry coefficient (r=0.46 and 0.43 for related sets of R and B lines). In this instance, information derived from DNA markers is substituting for the considerable time and records needed for traditional analysis of pedigrees and lineages. Also, RFLP data seem to portray genetic relationships more accurately than the methods based exclusively on the coancestry coefficient.

As with any method, DNA-based estimates of genetic diversity have inherent potential for error and bias. With DNA markers, there has been justifiable concern about laboratory technique, standards, and data interpretation. With a few precautions, errors may be minimized or available for further evaluation. Estimates based on DNA markers have an upward bias for f, but the bias may be a significant problem only for natural populations. Also, estimates of genetic similarity based on the number of DNA fragments in common between two individuals do not necessarily portray similarities based on common ancestry; the bands may merely reflect genes which are *identical in state* (i.e. alike in state) and not *identical by descent* (Smith and Smith, 1992). Despite these and perhaps other potential limitations, DNA markers have represented a significant improvement in plant breeders' perception of genetic diversity. On the basis of the number of available methods for detecting DNA polymorphism and relatively comprehensive coverage of the genome, DNA markers have become a standard tool for this aspect of plant-breeding programs.

FUTURE TRENDS AND OBJECTIVES

Integration of the sorghum genetic map with those of other species such as maize, rice, barley, and wheat should permit exchange and mutual exploitation of information and materials to expedite advancements in various aspects of basic biology and crop improvement. For example, several plant height mutants in maize have been tagged or cloned with transposable elements. DNA sequences of these genes could be used to isolate the related genes from sorghum. This could permit direct assessments and perhaps manipulation of QTL and important phenotypes. Likewise, the smaller genome of sorghum, its genetic map, and assemblage of valuable phenotypes and mutants could be used to facilitate and complement genetic investigations and advances in maize and other grasses.

BENEFITS FOR CROP IMPROVEMENT?

The benefits of comparative and integrated maps for plant breeding programs are substantial for the short and long term. These have been interspersed throughout the preceding sections and may be summarized as follows: (1) to the extent macromutations are utilized by breeders, integrated maps increase opportunities for indirect selection methods; (2) the ability to share genetic information between sexually isolated species should accelerate isolation of targeted genes; (3) the definition of crop gene pools should become broader and more precise for specific genes; (4) understanding of the biological basis of complex traits should improve by providing a common language to various branches of biology; (5) important genes may be localized by a variety of increasingly complementary methods; and (6) an element of objective hypothesis testing has become available for plant breeding.

The utility of comparative and integrated maps has some limitations. One limitation may relate to the observation that 10-20% of the low copy DNA clones from one species seem to be specific, or at least much more homologous to the source species (Fatokun et al., 1992; Pereira et al., 1994). That could represent a substantial number of genes. At least some of those genes might confer unique or neomorphic functions in the source species. Undoubtedly there will be many examples of species-specific low copy sequences that turn out to be very important genes. Nevertheless, there is so much to learn about so many shared sequences that comparative and integrative mapping is easily justified.

Recent developments and advances in several areas such as PCR-based detection methods of DNA polymorphisms, physical mapping, transformation, and informatics offer a wealth of genetic information, material, and opportunities for crop improvement programs. The first generation of DNA combines (i.e. thermal cyclers and sequencers) have been constructed and are being implemented. While they will not replace extant methods of plant breeding, they will, in certain circumstances provide considerable utility and advantages for the pursuit and understanding of genetic gain.

REFERENCES

Ahn, S., Anderson, J. A., Sorrells, M. A. and Tanksley, S. D. (1993). Homoeologous relationships of rice, wheat, and maize chromosomes. *Mol. Gen. Genet.* **241**, 483-490.

Ahnert, D., Austin, D., Lee, M., Livini, C., Openshaw, S. J., Smith, J. S. C., Porter, K. and Dalton, G. (1995). Genetic diversity among elite sorghum inbred lines assessed with DNA markers and pedigree information. *Crop Sci.* (in press).

Beavis, W. D., Grant, D., Albertsen, M. and Fincher, R. (1991). Quantitative trait loci for plant height in four maize populations and their associations with qualitative genetic loci. *Theoret. Appl. Genet.* **83**, 141-145.

Beavis, W. D., Lee, M., Grant, D., Hallauer, A. R., Owens, T., Katt, M. and Blair, D. (1992). The influence of random mating on recombination among RFLP loci. *Maize Genet. Coop. Newsl.* **66**, 52-53.

Bennetzen, J. L. and Freeling, M. (1993). Grasses as a single genetic system – genome composition, collinearity, and compatibility. *Trends Genet.* **9**, 259-261.

Bretting, P. K. and Widrlechner, M. P. (1995). Genetic markers and plant genetic resource management. *Plant Breeding Rev.* **13**, 11-86.

Chao, S., Baysdorfer, C., Heredia-Diaz, O., Musket, T., Xu, G. and Coe, E. H. Jr (1994). RFLP mapping of partially sequenced leaf cDNA clones in maize. *Theoret. Appl. Genet.* **88**, 717-721.

Devos, K. M., Millan, T. and Gale, M. D. (1993a). Comparative RFLP maps of the homoeologous group-2 chromosomes of wheat, rye, and barley. *Theoret. Appl. Genet.* **85**, 784-792.

Devos, K. M., Atkinson, M. D., Chinoy, C. N., Francis, H. A., Harcourt, R. L., Koeber, R. M. D., Liu, C. J., Masojc, P., Xie, D. X. and Gale, M. D. (1993b). Chromosomal rearrangements in the rye genome relative to that of wheat. *Theoret. Appl. Genet.* **85**, 673-680.

Dogget, H. (1988). *Sorghum.* Longmans, Green & Co, London.

Duncan, R. R., Bramel-Cox, P. J. and Miller, F. R. (1991). Contributions of introduced sorghum germplasm to hybrid development in the USA. In *Use of Plant Introductions in Cultivar Development*, part 1 (ed. H. L. Shands and

L. E. Weisner), pp. 69-101. CSSA Special Publication No. 17. Crop Science Society of America, Inc., Madison, WI, USA.

Fatokun, C. A., Menacio-Hautea, D. I., Danesh, D. and Young, N. D. (1992). Evidence for orthologous seed weight genes in cowpea and mung bean based on RFLP mapping. *Genetics* **132**, 841-846.

Grotewold, E. and Peterson, T. (1994). Isolation and characterization of a maize encoding chalcone flavonone isomerase. *Mol. Gen. Genet.* **242**, 1-8.

Helentjaris, T. (1993). Implications for conserved genomic structure among plant species. *Proc. Nat. Acad. Sci. USA* **90**, 8308-8309.

Lee, M. (1995). DNA markers and plant breeding programs. *Advan. Agron.* **55**, 265-343.

Melchinger, A. E. (1993). *Use of RFLP Markers for Analysis of Genetic Relationships among Breeding Materials and Prediction of Hybrid Performance* (ed. D. R. Buxton et al.), pp. 621-628. International Crop Science I (Proc. Int. Symp. Ames, IA, USA, 1992). Crop Science Society of America, Inc., Madison, WI, USA.

Pereira, M. G., Lee, M., Bramel-Cox, P., Woodman, W., Doebley, J. and Whitkus, R. (1994). Construction of an RFLP map in sorghum and comparative mapping in maize. *Genome* **37**, 236-243.

Pereira, M. G. and Lee, M. (1995). Identification of genomic regions affecting plant height in sorghum and maize. *Theor. Appl. Genet.* **90**, 380-388.

Rayapati, P. J., Portyanko, V. A. and Lee, M. (1994). Placement of loci for avenins and resistance to *Puccinia coronata* to a common linkage group in *Avena strigosa*. *Genome* **37**, 900-903.

Smith, O. S. and Smith, J. S. C. (1992). Fingerprinting crop varieties. *Advan. Agron.* **47**, 85-140.

Tanksley, S. D., Bernatzky, R., Lapitan, N. L. and Prince, J. P. (1988). Conservation of gene repertoire but not gene order in pepper and tomato. *Proc. Nat. Acad. Sci. USA* **85**, 6419-6423.

Tanksley, S. D., Ganal, M. W., Prince, J. P, De Vicente, M. C., Bonierbale, M. W. Broun, P., Fulton, T. M., Giovannoni, J. J., Grandillo, S., Martin, G. B. et al. (1992). High density molecular linkage maps of the tomato and potato genomes. *Genetics* **132**, 1141-1160.

Tanksley, S. D. (1993). Mapping polygenes. *Annu. Rev. Genet.* **27**, 205-233.

Teutonico, R. A. and Osborn, T. C. (1994). Mapping of RFLP and quantitative trait loci in *Brassica rapa* and comparison to the linkage maps of *B. napus*, *B. oleracea* and *Arabidopsis thaliani*. *Theor. Appl. Genet.* **89**, 885-893.

Veldboom, L. R., Lee, M. and Woodman, W. L. (1994). Molecular-marker-facilitated studies in an elite maize population. I. Linkage analysis and determination of QTL for morphological traits. *Theoret. Appl. Genet.* **88**, 7-16.

Weber, D. and Helentjaris, T. (1989). Mapping RFLP loci in maize using B-A translocations. *Genetics* **121**, 583-590.

Werner, J. E., Endo, T. R. and Gill, B. S. (1992). Toward a cytogenetically based physical map of the wheat genome. *Proc. Nat. Acad. Sci. USA* **89**, 11307-11311.

Printed in Great Britain © The Society for Experimental Biology 1996
SEB0027

QTL for insect resistance and drought tolerance in tropical maize: prospects for marker assisted selection

D. Hoisington[1], C. Jiang[1], M. Khairallah[1], J.-M. Ribaut[1], M. Bohn[2], A. Melchinger[2], M. Willcox[1] and D. González-de-León[1]

[1]CIMMYT, Int., Lisboa 27, Apdo Postal 6-641, Delg. Cuauhtemoc 06600, Mexico
[2]Institute of Plant Breeding, Seed Science and Population Genetics, University of Hohenheim, 70593 Stuttgart, Germany

SUMMARY

Insects and drought cause severe losses in the production of maize in many developing countries. Conventional breeding efforts to enhance the level of resistance to a number of insect pests and tolerance to drought have been successful, although only through large efforts of many breeders and over a large period of time. Continued improvements will only be possible through substantial investment of resources. Recently, success in identifying quantitative trait loci (QTL) in several plant species using various molecular marker systems offers alternative methods for accelerating conventional breeding programs. As the first step towards using molecular markers in CIMMYT's maize breeding program, restriction fragment length polymorphisms (RFLPs) have been used to understand the genetic basis of resistance to two corn borer species, southwestern corn borer and sugarcane borer, and to one major component of drought tolerance, anthesis-silking interval. A number of QTL with effects large enough to be regarded as significant in breeding were detected for each of these traits and many of them presented stable effects over environments. While variability in the number and location of QTL has been found when compared across populations, several loci were found to be quite consistent. Simple calculations can be made which estimate that the total genetic potential in maize for these traits is high. It is argued that to ultimately access and manipulate this potential, the use of linked molecular markers as indirect selectable markers is both feasible and necessary.

Key words: QTL, RFLP, Maize, Insect resistance, Drought tolerance, Plant height

INTRODUCTION

Insects and drought are two important causes of maize yield loss in developing countries (Mihm et al., 1989; Edmeades et al., 1992) and efforts to improve germplasm for both traits have been a major part of CIMMYT's Maize Program for many years. While improved varieties have been developed, breeding for resistance to insects and tolerance to drought has been laborious and time consuming. For insect resistance, large scale rearing facilities and repeated and effective artificial infestations are required at each cycle of selection (Smith et al., 1989). For drought tolerance, the yield loss under drought stress has been difficult to measure and has shown low heritability resulting in inefficient selection. While the anthesis-silking interval (ASI) has been found to be correlated with yield under drought stress (Bolaños and Edmeades, 1993), selection under drought stress requires two to three months without rain as well as adequate irrigation infrastructure to control the levels of the stress. Owing to these difficulties, we have performed several experiments with a view to using marker assisted selection (MAS) for improving the two traits.

For each experiment, a segregating population was derived from the cross of two lines clearly contrasting in the expression of the specific traits to be improved. We then obtained restriction fragment length polymorphism (RFLP) genetic linkage maps in order to identify genomic segments responsible for the expression of the traits. These traits are under polygenic control and the identified genomic segments are known as quantitative trait loci (QTL). Experiments were then designed for the effective transfer, through MAS, of these QTL from a donor resistant/tolerant line into an elite recurrent line requiring improvement of the character in question.

In this report we present results from some of our experiments with an emphasis on the information required to assess the possible progress that could be achieved by MAS. Also, by comparing the results across different experiments, we discuss the underlying genetic structure of several traits and the consequences thereof for the effective application of MAS in CIMMYT maize germplasm. In addition to insect resistance and ASI, results about plant height are also presented.

EXPERIMENTAL DESIGN AND GENETIC MAPPING

Further details of the three experiments summarized here are presented by Khairallah et al. (1996), Bohn et al. (1996) and Ribaut et al. (1996). Briefly, each experiment was based on a specific cross between two carefully selected parents having

Table 1. Brief description of three experiments for mapping QTL of SWCB and SCB resistance and ASI

Cross $P_1 \times P_2$	Description of parents	Number of individuals (and markers)	Total map genetic length (and average interval size) (cM)	Field evaluation*
(1) Ki3 (P_1) × CML139 (P_2)	P_1: Tropical late, susceptible to SWCB and SCB	475 (128)	1968 (17)	Tlaltizapan 90B, 91A 92A; Poza Rica 90B
	P_2: Subtropical intermediate, resistant to SWCB and SCB			
(2) CML131 (P_1) × CML67 (P_2)	P_1: Subtropical intermediate, very susceptible to SWCB and SCB	190 (100)	1535 (17)	Tlaltizapan 92A, 93A; Poza Rica 93A
	P_2: Tropical late, very resistant to SWCB and SCB			
(3) Ac7729/TZSRW (P_1) × Ac7643 (P_2)	P_1: Tropical late, short ASI	234 (142)	1760 (12.5)	Tlaltizapan 94A
	P_2: Tropical late, long ASI			

*(A) Winter dry season; (B) Summer rainy season. Each trial had two replicates.

large differences for the traits investigated. Two of them were designed for the study of the inheritance of resistance to southwestern corn borer (SWCB, *Diatraea grandiosella*) and sugarcane borer (SCB, *Diatraea saccharalis*) and, the third, the anthesis-silking interval (ASI). A brief description of the three experiments is given in Table 1. Genetic mapping was based on the F2 population derived from one or two self-fertilized F1 hybrid cobs from crosses between inbred lines. DNA samples were from F2 individual plants and phenotypic data were based on evaluation of the F2:3 families (sib-mated F3 families derived from the selfed F2 individuals) grown in one or two CIMMYT experiment stations for two or three years. For the evaluation of SWCB and SCB resistance, 10 plants were artificially infested with neonate larvae and the leaf feeding damage per plant was graded using a rating scale from 1 (no visible damage) to 10 (dead heart and all leaves with long lesions) (Davis and Williams, 1989). Family means were used for the statistical analysis. For ASI, two stress conditions, intermediate and severe, were applied to the F2:3 families by controlling the irrigation before and during the flowering period. The number of days of ASI was recorded for each plant and the family means were used in the analysis. Of the various other important traits that were measured in F2:3 families, we have only included the data obtained for plant height as an example of QTL comparisons that can be made across several segregating populations.

The RFLP linkage maps were constructed independently for each of the three experiments using the program *Mapmaker* (Lander et al., 1987). In order to summarize the results for all experiments, an integrated map is needed. There were 58 loci in common between crosses 1 and 2 (about half the total number of loci), and 45 and 36 loci between crosses 1 and 3, and 2 and 3, respectively (about one third of the total number of loci). The greater proportion of common markers between crosses 1 and 2 may indicate the greater similarity between the parents chosen for the insect resistance study. The selection of parents for cross 3 was made in other source germplasm and for very different reasons (differences in ASI). There were 27 marker loci common to all three crosses, which is about one-fifth of the average number of marker loci in each population. The order of the common markers was consistent over the three maps as well as with published maize maps (Burr et al., 1988; Gardiner et al., 1993), except for three pairs of closely linked markers (with recombination frequency less than 20 cM) in cross 2, which were flipped. The common marker loci will be used here to identify the location of common QTL in different crosses. In order to compare QTL positions in the large gaps between the common markers, we first tested the differences in interval sizes between the common markers in different crosses. Although the differences appeared to be large, when the large sample variance of the maximum likelihood estimates for recombination frequency was used to do the testing, only 3 out of 16 intervals were observed to be barely significant at 0.01 levels, and for 2 of these, segregation distortion for one or both markers was also observed. Although we know that recombination frequency may be affected by many factors, such as interference, it is likely that the differences observed here were due mainly to sampling error and, in some cases, segregation distortion. Thus, a combined map was produced for crosses 1 and 2 by treating data on non-common markers as missing in the population; the resulting map was used to identify chromosome segments of interest across both crosses. Finally, for comparing positions in cross 3 with that of crosses 1 and 2, relative distances from the nearest common marker were calculated.

QTL MAPPING

Statistical methods

Several methods, such as single-factor analysis of variance and interval mapping by *Mapmaker/QTL* (Lander and Botstein, 1989) have been used for QTL mapping in the three crosses. However, the results presented here were mainly obtained using the composite interval mapping (CIM) procedure developed more recently (Zeng, 1994). This method integrates interval mapping with regression by using markers as cofactors. By using cofactors it minimizes the effects from other possible QTL while one QTL is tested, thus increasing the power of QTL detection and reducing the bias of estimates of QTL effects and positions. However, in most cases, the results obtained with *Mapmaker/QTL* were confirmed by CIM, except that the latter method was more effective in resolving and providing estimates for multiple linked QTL.

Initially, CIM was performed using unlinked markers as cofactors for QTL detection, thus maximizing the power of detection of QTL segregating independently. Then, linked markers with different recombination frequencies, if available, were selected and added as cofactors to separate possible linked QTL and to refine the estimation of QTL effects and positions. It is expected and also shown by our results that multiple linked QTL are quite common for quantitative traits even though many of them may be too small to be recognized.

QTL detecting power was further increased by joint analysis of multiple trials in different locations and years (Jiang and Zeng, 1995). In effect, similar (but not significant) tendencies in different trials may help detect additional QTL, as well as improve the estimation of the position and effects of other previously defined QTL. Finally, possible QTL could be tested by environment interactions resulting from estimates of QTL effects in different trials of a given cross.

The significance level used for QTL detection with CIM was fixed at likelihood ratios of 13.8 for individual trials, and 17.8 and 21.4 for joint analyses of two and three trials, respectively; in the case of SWCB resistance in cross 2, likelihood ratios of 10.5, 15.3 and 18.6 were used owing to smaller sample size and number of replicates. Whenever an individual or a joint analysis was significant in a given interval, a QTL was declared. QTL by environment interactions were tested by using the chi-square test with a significance level $\alpha=0.05$ representing the experimentwise error rate. The significance level for each comparison was then calculated as $\alpha'=\alpha/k$, where k denotes the total number of QTL detected for the trait in that experiment.

In order to discover possible common QTL between different crosses, genetic variation at the estimated QTL position is needed. Based on computer simulations (data not shown) and considering the sample size, marker map density and the QTL effects in the experiments, it was possible to establish that the error in estimates of a QTL position would be around 15 to 30 cM. Thus, when two QTL were identified in different crosses and their positions fall within that error interval, they were considered as one and the same. In any case, it is clear that the discrimination of linked QTL within 15 to 30 cM of each other would be difficult with current methods (including CIM) with the population sizes used in these experiments. If two QTL independently estimated from different crosses have their position within this interval, we will recognize them as one QTL. Actually, separation of closely linked QTL with such distance would be difficult even by CIM and by conventional mass selection in breeding populations.

QTL mapping for insect resistance

The leaf feeding damage ratings of the parental lines of crosses 1 and 2 under infestation and the QTL detected for SWCB and SCB resistance are summarized in Table 2.

The difference in average leaf damage rating between the two parents of cross 2 (4.0) is double that observed for cross 1 (2.0). It was possible to detect 10 QTL for SWCB resistance in cross 1, but only 5 and 8 in cross 2 for SWCB and SCB resistance, respectively. This may be partly due to the greater number of $F_{2:3}$ families in cross 1 ($n=476$) than in cross 2 ($n=171$), as well as to intrinsic differences in the germplasm itself. The total resistant allelic effects of the QTL detected from the resistant parent can explain about three fourths of the parental difference in cross 1 and about one half in cross 2. The absolute value of the total effects in cross 2, however, is larger than that in cross 1 due to the greater difference in the parental resistance of cross 2. These QTL are distributed over almost all chromosomes.

For most of the detected QTL, the estimated effects were quite consistent over environments and only four revealed significant QTL by environment interactions. In total, six out of 23 QTL changed the sign of their estimated additive effects and thus contributed to a decrease in the leaf ratings in one trial but to an increase in another.

The QTL with the largest effect was detected in cross 1 on chromosome 6, in the trial conducted in CIMMYT's Tlaltizapan Experiment Station. This QTL also had the most significant QTL by environment interaction, and is thought to be due to the high iron deficiency in the plot where the trial was conducted. Moreover, QTL were detected for all other traits measured in that trial at the same genomic position, thus illustrating the perils of interpretation of QTL data for a single trait, as well as some aspects of QTL by environment interactions.

To compare the QTL distribution of different experiments, all QTL detected were plotted on the joint map (Fig. 1), positioned relative to markers common to the two mapping populations. The position of four pairs of QTL from two experiments appeared to match in the same chromosome segments: one in chromosome 2, one in chromosome 7, and two in chromosome 9. Other QTL were scattered in different chromosome regions. However, all common QTL were between SWCB and SCB resistance and none of the QTL for SWCB resistance in cross 1 and 2 were in common. One possible explanation for this is the low QTL detecting power for SWCB resistance in cross 2.

Finally, the heritability of the family mean leaf damage rating over two replicates was 0.19 for SWCB in cross 1 (which is rather low due to the large genotype by environment interaction in the trial with iron deficiency), 0.47 and 0.38 for

Table 2. QTL detected for insect resistance, ASI and plant height in three experiments

Cross	Trait	Trait value		Total QTL	Allelic effects*		Chromosomes
		P_1	P_2		P_1	P_2	
1	SWCB	7.4±0.1	5.4±0.1	10	1.6 (2.3)†		3,5,6,7,8,9
2	SWCB	8.4±0.2	4.3±0.2	5	2.2		1,7,9
2	SCB	8.3±0.2	4.6±0.1	8	2.0		1,2,5,7,9
1	ASI	0.6±0.1	1.9±0.1	9	−1.8	−2.0	1,3,5,6,7,8,9,10
3	ASI	−0.3±0.3	7.0±0.7	7	−5.3	−2.7	1,2,3,6,8,10
1	PltHt	136±1.0	113±1.8	11	21.3	54.8	1 ~ 10
2	PltHt	109±1.4	77±1.4	7	36.1	8.6	2,3,4,5,7,8
3	PltHt	129±1.5	147±2.5	8	19.6	39.1	1,3,4,5,7,9,10

*Total effects of the alleles from each parent averaged over trials and multiplied by 2 to make it equivalent to the effect of a homozygote.
†Due to iron deficiency in trial 1 of cross 1, 2.3 is the average for all three trials and 1.6 is the average for trials 2 and 3.

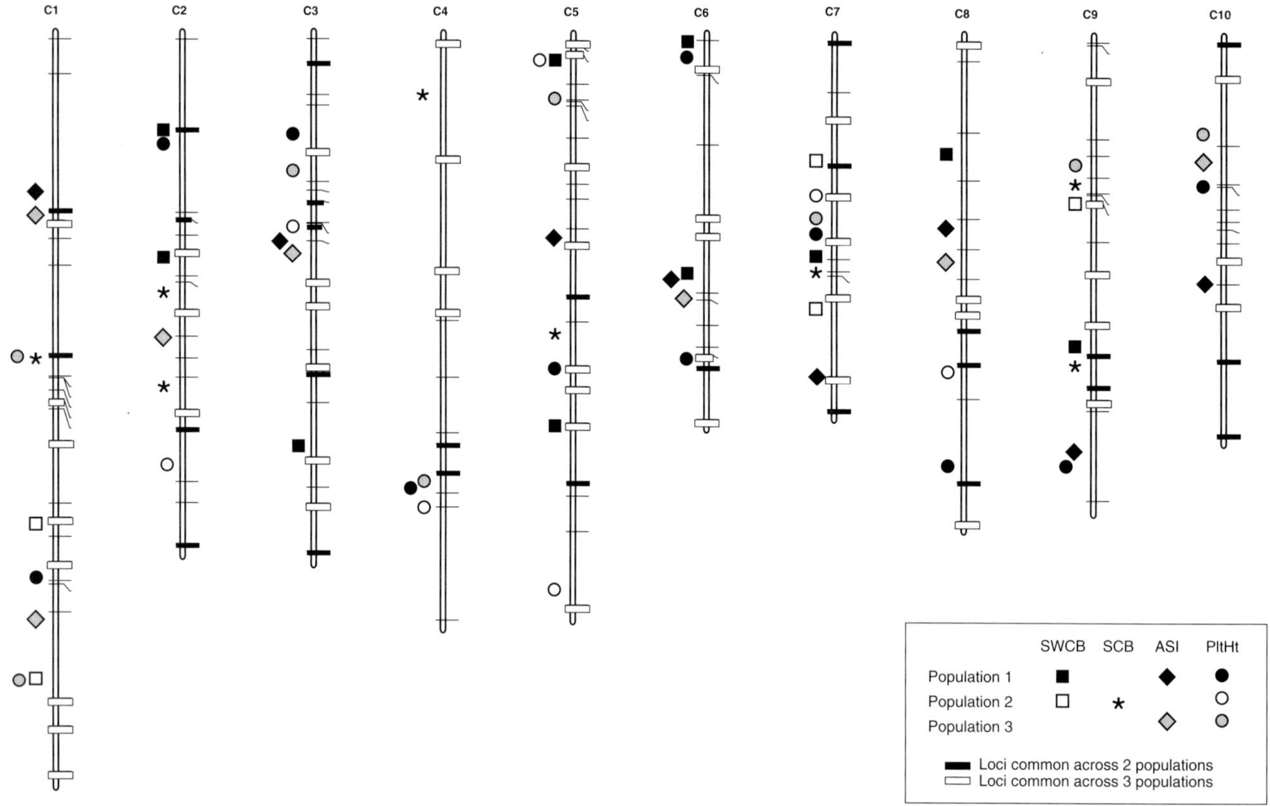

Fig. 1. RFLP linkage map of maize with positions of QTL for southwestern corn borer (SWCB), sugarcane borer (SCB), anthesis-silking interval (ASI) and plant height (PltHt) across three F_2 populations.

SWCB and SCB resistance, respectively, in cross 2. Given the magnitude of these estimates and the absence of major genes with predominant effects, it is not surprising to have obtained the results detailed above.

QTL mapping of the anthesis-silking interval (ASI)

While the effect of drought on yield reduction may be a very complex trait to measure and analyze, ASI can be measured with relatively low error. Cross 3 was specifically designed for mapping QTL responsible for ASI, given a difference in ASI of about 7 days between the two selected parents. The heritability of $F_{2:3}$ family means over two replicates for the trait was 0.55 in cross 3, and 0.65 for the average over two stress conditions.

A total of 8 QTL were detected and are shown in Fig. 1. The drought tolerant parent (P_1) had desired ('short ASI') alleles at 5 loci while the susceptible parent (P_2) had desired alleles at 2 loci. This is reflected in the clear transgressive segregation observed in this population, some individuals having more 'short-ASI' alleles than P_1. At one QTL on chromosome 1 the estimated allelic effect changed sign between intermediate and severe stress conditions. The effects of the 4 most desired alleles from P_1 were, on average, twice as large as the two from P_2. The total effect of the four alleles added up to 5.3 days for a homozygous line. They were distributed on chromosomes 2, 6, 8 and 10. The QTL by environment interactions at these four loci were not significant, which means that the expression of these QTL was quite stable over environments. The effect

should reach 8.0 days for individuals homozygous for all the short-ASI alleles from both parents.

Although cross 1 was not designed for mapping QTL of ASI, this variable was also recorded in the field and analyzed. This experiment was run under full irrigation. The average difference of ASI between the two parents was only 1.3 days and the genetic variance in the population was low with a heritability of 0.34 for the family means over two replicates. However, due to the large number of families measured (475), 9 QTL were detected. Alleles for short ASI at the detected QTL were distributed between the two parents with 3 for P_1 and 5 for P_2 with a total allelic effect of 1.8 and 2 days, respectively. Furthermore, the estimated effects were all close to 0.5 days for a homozygote at all 9 regions. These QTL effects are much smaller than those estimated for cross 3. This implies that the short-ASI parent of cross 3 (P_1) did indeed include several good alleles with relatively large effects for short ASI and thus showed drought tolerance. By comparing the distribution of the QTL in crosses 1 and 3, it is found that four QTL pairs have very close positions on chromosomes 1, 3, 6 and 8; these may be regarded as QTL in common between the distinct germplasm sources that were used for the two crosses.

QTL mapping for plant height

In none of the three crosses was plant height specifically considered in selecting the parental lines. However, differences between parents did exist. Given the diversity of the parental lines, it is expected that both parents of a given cross would be likely to contribute alleles with positive effects on plant height. Therefore, the segregation for plant height in each population would be much larger than the differences between two parents. This expectation is borne out by the observed results listed in Table 2.

The differences in the number of QTL detected in the three crosses may be a reflection of differences in sample size: cross 1 had the largest and cross 2 the smallest sample; 11 QTL were detected for cross 1, and only 7 and 8 for crosses 2 and 3, respectively; the total effects of the QTL detected in crosses 1, 2 and 3 were 76 cm, 45 cm and 58 cm, respectively.

The plant height QTL were distributed over all ten chromosomes. Only three out of a total of 26 changed the sign of their estimated additive effects over trials. The largest QTL effect averaged over environments in a homozygote for all 23 QTL (with estimated effects in the same direction over environments) is 13.2 cm, and the smallest is 2.8 cm. Therefore none of the QTL can be regarded as major genes in the sense of classical breeding and they were all polygenes.

QTL by environment interactions were significant only at 3 QTL, of which the largest one was very probably due to iron deficiency in the field plot as for insect resistance. The difference between the two environments for a homozygote for this QTL was 22 cm. The estimated QTL effects in cross 1 may be worth special notice. Although P_2 is shorter than P_1 due to the large genotype by environment interaction, it still provides more QTL with effects of increasing plant height.

Comparing the QTL maps, two segments were found to have QTL for all three crosses: one was on chromosome 4 at 7 cM, 15 cM and 5 cM from *umc15a*; the other one was on chromosome 7 at 28 cM, 12 cM and 20 cM from *umc116a*. These distances were scaled relative to the joint map of crosses 1 and 2. There were three more pairs of QTL that appeared to have close positions on chromosomes 3, 5 and 10.

PROSPECTS FOR MARKER ASSISTED SELECTION

MAS by using desired inbred lines as parents

Successful applications of QTL mapping through marker assisted selection (MAS) in plant breeding depend on the precision of both the detection of the QTL and the estimation of their effects and positions on the chromosomes. Our results are encouraging. For both insect resistance and drought tolerance, 4 to 7 QTL with major effects in each cross were detected and can explain more than 50% of the genetic difference between parental lines from widely different germplasm backgrounds. None of the QTL detected had predominant effects on the trait thus suggesting that the two traits are typical quantitative traits. Without epistatic effects, the accumulation of these QTL would be expected to provide considerable improvement for the trait in any background. Most of the QTL detected were consistent in their effects over locations and years even though the family by environment interactions were significant especially for insect resistance. It is clear that differences between the QTL maps for each trait do exist. ASI has higher heritability and fewer QTL with relatively larger effects compared to insect resistance. Thus, ASI may be regarded as a genetically more simple trait than insect resistance. It is expected that the application of MAS to improve ASI should be more successful in the short term than improving insect resistance in populations derived from the three crosses investigated.

'Genetic potential' in tropical maize germplasm

In all three experiments, the genetic variation unexplained by detected QTL remained large, especially for insect resistance. Also, most of the QTL detected in different crosses were different for all three traits analyzed. This implies that different sets of QTL would each be responsible for specific trait expression in different crosses. We do not and cannot yet know how many potential QTL may exist in tropical maize germplasm for any given trait, but a simple calculation may give us a rough idea. Based on a dense marker map (our genetic maps with average interval size 15 cM would be regarded as relatively dense), it can be postulated that the fraction of markers in common between crosses are expected to be directly proportional to the fraction of QTL in common. But the observed correlation between two fractions would depend on the number of potentially segregating QTL in the germplasm. If only a few QTL with relatively large effect are responsible for the segregation of a trait in the germplasm, different crosses would have more chance to detect common QTL. By contrast, if numerous QTL, each of which has a relatively small effect, segregate in the germplasm, the chance of detecting common QTL in different crosses would be less due to sampling error. Thus, a total of 15 QTL were detected in crosses 1 and 2 for SWCB resistance, with only one pair mapping to a similar position; for ASI, on average four out of eight QTL were in common between crosses 3 and 1; finally, for plant height, on average three out of nine QTL were in common between all crosses.

Now, let θ denote the proportion of probes revealing monomorphic loci in one cross; then, the proportion of polymorphic loci would be $(1-\theta)$. The average proportion of polymorphic loci in common between two random crosses will be $(1-\theta)^2$. If one assumes that the fraction of molecular loci in common between two crosses is proportional to the number of common QTL, then the proportion of 'new' QTL detectable in a second population would be $1-(1-\theta)$, that is θ. Then, $(1-\theta)$ can be estimated by the ratio of the number of common markers to the total number of markers in each cross. In our experiment, it was 0.4 on average (this is much smaller than the actual proportion of polymorphic markers among all probes tested, that is around 0.7, due to the selection of parents with large genetic differences).

The fraction of common QTL in different crosses for ASI and plant height in our experiment are quite close to this ratio. This implies that the number of QTL with an effect large enough to be detected by an experiment of the size used here would have a lower bound of around 21. The observed number of plant height QTL over the three crosses was 19, which fits well with the expected number. But considering the sampling error and the QTL with relatively small effects, this number is a minimum. In the case of insect resistance, the minimum number of QTL would be considerably larger. In addition, the genetic resource may be even richer owing to the existence of multiple alleles at those QTL.

Although the resistant and tolerant parents in our experiments have been derived from many cycles of selection during the improvement of the germplasm pool, further improvement is possible through the accumulation of more desirable alleles at a multiplicity of QTL. MAS should prove instrumental in such an endeavor.

REFERENCES

Bohn, M., Khairallah, M. M., González-de-León, D., Hoisington, D. A., Utz, H. F., Deutsch, J. A., Jewell, D. C., Mihm, J. A. and Melchinger, A. E. (1996). QTL mapping in tropical maize: I. Genomic regions affecting leaf feeding resistance to sugarcane borer and other traits. *Crop Sci.* (in press).

Bolaños, J. and Edmeades, G. O. (1993). Eight cycles of selection for drought tolerance in lowland tropical maize. II. Responses in reproductive behavior. *Field Crops Res.* **31**, 253-268.

Burr, B., Burr, F. A., Thompson, K. H., Albertson, M. C. and Stuber, C. W. (1988). Gene mapping with recombinant inbreds in maize. *Genetics* **118**, 519-526.

Davis, F. M. and Williams, W. P. (1989). Methods used to screen maize for and to determine mechanisms of resistance to the southwestern corn borer and fall armyworm. In *Toward Insect Resistant Maize for the Third World: Proceedings of the International Symposium on Methodologies for Developing Host Plant Resistance to Maize Insects* (ed. J. A. Mihm, B. R. Wiseman and F. M. Davis), pp. 101-108. CIMMYT, Mexico, DF.

Edmeades, G. O., Bolaños, J. and Lafitte, H. R. (1992). Progress in breeding for drought tolerance in maize. In *47th Annual Corn & Sorghum Research Conference* (ed. D. Wilkinson), pp. 93-111. ASTA Proceedings.

Gardiner, J. M., Coe, E. H., Melia-Hancock, S., Hoisington, D. A. and Chao, S. (1993). Development of a core RFLP map in maize using an immortalized F2 population. *Genetics* **134**, 917-930.

Jiang, C. and Zeng, Z.-B. (1995). Multiple trait analysis of genetic mapping for quantitative trait loci. *Genetics* **140**, 1111-1127.

Khairallah, M. M., Bohn, M., Jiang, C., Deutsch, J. A., Jewell, D. C., Mihm, J. A., Melchinger, A. E., González-de-León, D. and Hoisington, D. A. (1996). Molecular mapping of quantitative trait loci for southwestern corn borer resistance in tropical maize (in preparation).

Lander, E. S., Green, P., Abrahamson, J., Barlow, A., Daly, M. J., Lincoln, S. E. and Newburg, L. (1987). MAPMAKER: an interactive computer package for constructing primary genetic linkage maps of experimental and natural populations. *Genomics* **1**, 174-181.

Lander, E. S. and Botstein, D. (1989). Mapping mendelian factors underlying quantitative traits using RFLP linkage maps. *Genetics* **121**, 185-199.

Mihm, J. A., Wiseman, B. R. and Davis, F. M. (editors) (1989). *Toward Insect Resistant Maize for the Third World: Proceedings of the International Symposium on Methodologies for Developing Host Plant Resistance to Maize Insects.* CIMMYT, Mexico, DF.

Ribaut, J.-M., Hoisington, D. A., Deutsch, J. A., Jiang, C. and González-de-León, D. (1996). Identification of quantitative trait loci under drought conditions in tropical maize. I. Flowering parameters and the anthesis-silking interval. TAG. **92**, 905-914.

Smith, M. E., Mihm, J. A. and Jewell, D. C. (1989). Breeding for multiple resistance to temperate, subtropical, and tropical maize insect pests at CIMMYT. In *Toward Insect Resistant Maize for the Third World: Proceedings of the International Symposium on Methodologies for Developing Host Plant Resistance to Maize Insects* (ed. J. A. Mihm, B. R. Wiseman and F. M. Davis), pp. 222-234. CIMMYT, Mexico, DF.

Zeng, Z.-B. (1994). Precision mapping of quantitative trait loci. *Genetics* **136**, 1457-1468.

Printed in Great Britain © The Society for Experimental Biology 1996
SEB0028

The nucleotype, the natural karyotype and the ancestral genome

Michael D. Bennett

Jodrell Laboratory, Royal Botanic Gardens, Kew, Richmond, Surrey, TW9 3DS, UK

SUMMARY

New knowledge of synteny and collinearity promises to unify genetics and to affect our perception of higher order genome structure. This exciting new synthetic approach emphasizes genomic similarities rather than diversity. Two other aspects of genomic form and organisation, offering potentially unifying concepts in genome studies are: the nucleotype, and the natural karyotype. Genome size varies greatly between eukaryotes, and shows many strikingly precise correlations with phenotypic characters, independent of information encoded in DNA. Such nucleotypic correlations, based on biophysical absolutes, apply to all species, irrespective of genome size or chromosome number, and set limits on the range of phenotypes which can be expressed by genic control. Thus, knowledge of nucleotypic effects has considerable predictive value which can help to unify our understanding of genomes. Other studies of reconstructed nuclei have shown that: (1) the basic haploid genome exists as a real structural unit in nuclear architecture; while (2) the mean spatial arrangement of its heterologues also exists as a natural karyotype which is predictable using a simple model. Recently reported conceptual alignments of the maize genomes, which reflect the circularized ancestral grass genome, show interesting similarities with the orders of centromeres in their natural karyotypes predicted by the Bennett model. The basis of this phenomenon (if repeated in other species), and of selection which retains the ancestral genome form despite changes in basic chromosome number, may need to be explained. Perhaps the overall 3-D structure of the genome has some critical functional significance, essential for development. If so, a knowledge of this common structure would further unify our understanding of genomes and their evolution.

Key words: Nucleotype, Natural karyotype, Ancestral genome

INTRODUCTION

The 1990s will probably be remembered as the decade of biodiversity, when the need to conserve the fullest possible range of expressions of genomic variation was first widely grasped, and enshrined in the Biodiversity Convention in 1992. Yet it may also be seen (after the 1950s) as a second decade of biounification if, as seems likely, new knowledge of widely disparate genomes converges strongly to produce a much more unified framework of understanding. Genomes are already unified by a common genetic material (DNA) and by a common genetic code. Now, as described elsewhere in this volume, knowledge of highly conserved gene orders (synteny and collinearity) promises to unify genetics at a different level, affecting our perception of higher order genome structure. It remains to be seen just how far, including over what taxonomic distance, detectable and useful synteny and collinearity extends. Already it promises to abolish many old distinctions between wheat, maize, and rice genetics, and to produce a new unified cereal genetics (Moore et al., 1995a,b).

This exciting new synthetic approach emphasizes genomic similarities rather than diversity. In this context, two other aspects of genomic form and organisation (both originally defined at the Plant Breeding Institute, Cambridge) are worth considering as further potentially useful unifying concepts in genome studies, namely: the nucleotype, and, the natural karyotype.

THE NUCLEOTYPE

Genome size varies enormously between eukaryotes of similar genetic and organismic complexity. For example, DNA amount in the unreplicated haploid nuclear genome (the 1C value) of angiosperms varies over 1,000-fold, from less than 0.2 pg in *Arabidopsis thaliana* to 127 pg in *Fritillaria assyriaca* (Fig. 1) (Bennett and Smith, 1976; Bennett and Leitch, 1995).

Comparative studies of angiosperms have played a leading role in showing that DNA C-value is correlated with a wide range of phenotypic characters at the cellular level (Van't Hof and Sparrow, 1963; Bennett, 1973, 1976, 1985, 1987), including many size and mass characters, and rates of development. For example, Fig. 2c compares genome size with the duration of meiosis at 20°C in 18 diploid species. Such relationships seen individually are often strikingly close, but when several plots for widely different characters are all viewed together (Fig. 2a-d), this impression is strongly reinforced (Bennett, 1985, 1987). Clearly DNA amount correlates closely with many important phenotypic characters, and many of the relationships between DNA amount and various cellular characters are strikingly precise for biological phenomena, being more reminiscent of physical or chemical relationships.

It is impossible to increase DNA C-value greatly without also increasing the minimum chromosome, nuclear and cell volume (e.g. Fig. 1 shows the impossibility of containing the nucleus, the chromosomes, and even the DNA of *Fritillaria assyriaca* in the small cells of *Arabidopsis thaliana*). This fact is quite independent of the base sequence of the additional DNA, and of whether it is genic or not. Thus, nuclear DNA influences the phenotype in two ways, *first* by the expression of its genic content, and *second* by the physical effects of its mass and volume, which impose absolute limits on the range of phenotypes which can be expressed by genic control. Bennett (1971, 1972) coined the term 'nucleotype' to define those conditions of the nuclear DNA which affect the phenotype independently of its encoded informational content. Many of the correlations known between variation in DNA C-value and cell size, volume and mass and rate of development are largely or partly nucleotypic in origin, and to that extent, are causal in nature. Much evidence that nucleotypic effects at the cellular level are additive, and extend to the size, mass, development and distribution of multicellular structures, has also been found (Bennett, 1972, 1973, 1976, 1987).

One importance of nucleotypic correlations based on bio-physical absolutes is that they apply to all species, irrespective of genome size or chromosome number, setting limits on the range of phenotypes which can be expressed by genic control. As such they can help considerably to unify our understanding of genomes. Knowledge of genome size has considerable predictive value. For example, the duration of meiosis in *Ara-*

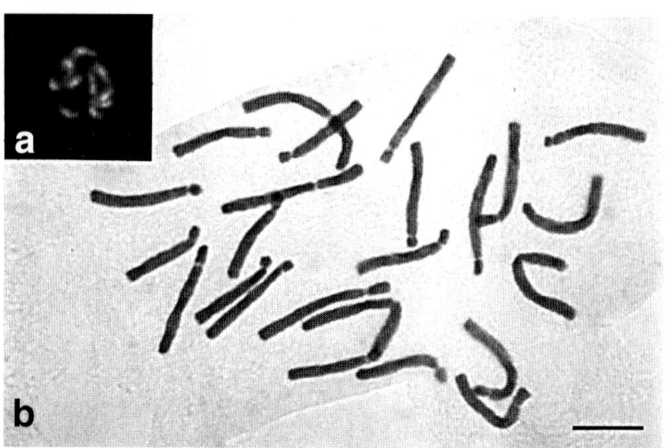

Fig. 1. Mitotic metaphase chromosomes of (a) *Arabidopsis thaliana* (1C DNA = 0.2 pg), and (b) *Fritillaria assyriaca* (1C DNA = 127 pg). Bar, 5 μm. Fig. 1a was kindly supplied by Professor J. S. Heslop-Harrison.

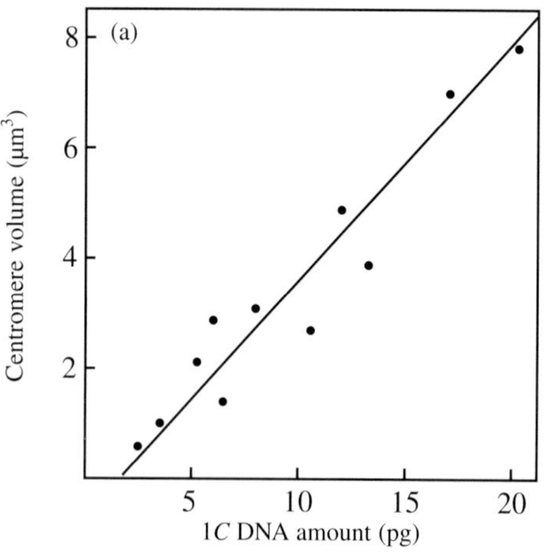

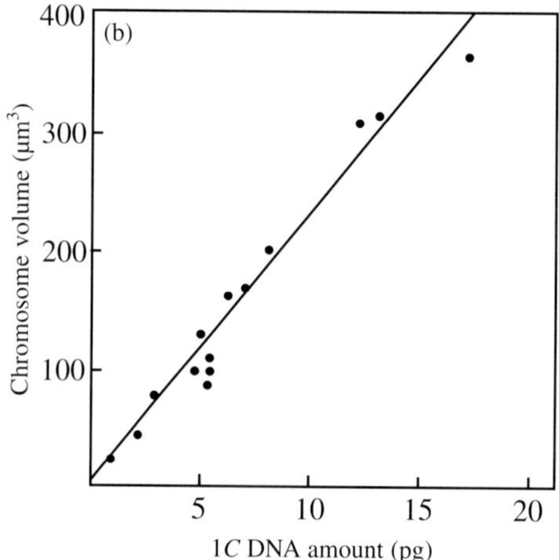

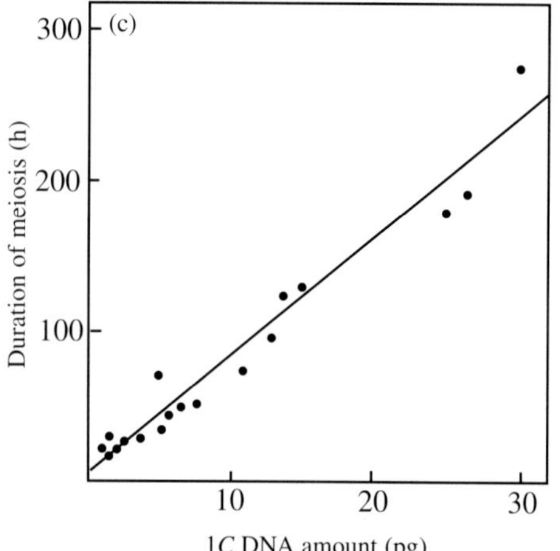

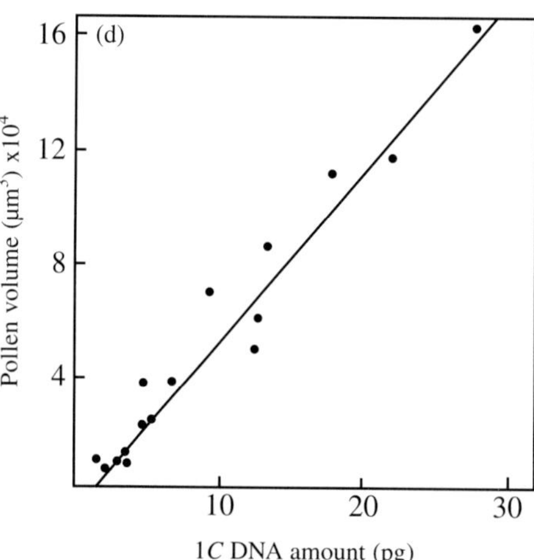

Fig. 2. Relationships in angiosperms between 1C DNA amount and (a) total volume of centromeres per nucleus in 11 species (from Bennett et al., 1981); (b) total mitotic metaphase volume per cell in 14 species (from Bennett et al., 1983); (c) duration of meiosis in 18 diploid species grown at 20°C (from Bennett, 1977); and (d) pollen grain volume at anther dehiscence in 16 species of wind-pollinated grasses (from Bennett, 1972).

bidopsis thaliana has not been estimated experimentally, as far as the author is aware. However, the regression shown in Fig. 2c predicts its duration to be about 10 hours at 20°C. Moreover, in species with very large genomes, the duration of meiosis alone is similar to that of the entire minimum generation time (about 4 weeks) for species with very small genomes such as *Arabidopsis thaliana*. From this evidence alone one can predict (Bennett, 1972), that all species which can complete a life cycle in under 6 weeks (known as 'ephemerals') must have very small genomes. Twenty-three years later, genome size data are now known for 2,500 angiosperm species (Bennett and Leitch, 1995), but no exception to this prediction has been found. Thus, knowledge of genome size and understanding of nucleotypic effects has considerable predictive value which can help to unify our understanding of genomes.

So far, in dealing with variation in genome size and its consequences, the genome has been considered as a unit. However the genome is divided between several chromosomes, which raises the question of how variation in DNA C-value is distributed between them in speciation.

Several detailed studies have shown that variation in DNA C-value between related species is often non-random and maintains chromosome or chromosome arm rankings. Differences in C-value involved either a constant increment, or a constant proportional difference, affecting every chromosome (Kenton et al., 1990). Clearly there are constraints and controls limiting the amount and distribution of DNA gain or loss in the karyotype, including mechanisms which can operate to maintain overall karyotypic form at the level of chromosome arm size (Rees, 1984). 'What determines these forms of karyotype orthoselection?' To suggest an answer we shall consider a second concept, that of 'the natural karyotype'.

THE NATURAL KARYOTYPE

Work started in Cambridge in the late 1970s to investigate the spatial arrangement of mitotic metaphase chromosomes in several grass species and hybrids, all with bi-armed chromosomes. Electron microscope, serial-section 3-D reconstruction techniques were used to study over 500 physically undistorted cells. This revealed striking evidence of non-random genome and chromosome arrangements (Bennett, 1982, 1988).

These studies showed first, that there is a highly significant tendency for haploid genomes to be spatially separate in both interspecific diploid hybrids, and in diploid species. In hybrids the tendency was for a concentric or radial separation of parental genomes (Finch et al., 1981; Schwarzacher-Robinson et al., 1987; Leitch et al., 1991; Schwarzacher et al., 1992), while in the diploid species there was a strong tendency for side-by-side separation of two haploid genomes (Bennett, 1982, 1983, 1984b). Such results were interpreted as showing that the haploid genome is a basic structural unit in nuclear architecture (Bennett, 1988). In other words, the set of chromosomes portrayed together on paper in a karyogram, does tend to be spatially organised as a real unit in vivo.

Second our work showed repeatedly that the spatial arrangement of different chromosome types (i.e. heterologues) within a separate basic haploid set is non-random. Indeed, the mean spatial order of heterologues is predictable using a simple model which orders a complete simple haploid genome so that each chromosome is associated with two constant neighbours. In essence it is based on associating pairs of most similar long and pairs of most similar short, chromosome arms throughout the karyotype, except at one point (termed the 'discontinuity' or the 'ends'), to minimize the total sum of differences between associated arms for the genome (Bennett, 1982).

Bennett (1983) listed the following rules for predicting the mean spatial order of chromosomes in a simple haploid genome, known as 'the Bennett model': (1) obtain a very accurate karyotype. (2) Separate long and short arms into two groups. (3) Rank arms within each group in decreasing size order. (4) Pair up arms of most similar length within each group. (5) Check that the result of pairing is permissible. NB, a permissible model uses every chromosome once, to produce a single unbroken chain. (6) For a permissible chain, calculate the size difference (d) between each pair of paired-up arms. (7) Calculate Σd, i.e. sum all values of d except that for the 2 arms at the discontinuity. (8) Check that no other permissible arrangement of arms gives a smaller value of Σd. (9) The permissible arrangement of arms giving the smallest value for Σd is the most likely prediction of their mean spatial order.

Fig. 3a shows the haploid complement of diploid *Secale africanum*. All seven chromosomes are readily identified in single cells using relative size, arm ratio and the presence and size of telomeric heterochromatic segments. Fig. 3b shows their predicted order using the model, namely: 1-2-4-3-5-6-7. Seven chromosomes, the basic number in *Secale* can be arranged in a closed chain in 360 different ways, ignoring the 360 mirror images. We fed the 3-D coordinates and identity of each centromere of a haploid genome of reconstructed cells into the computer which then calculated for each of the 360 orders the distance travelled in passing once through each centromere while starting and finishing at the same centromere. It then ranked the 360 orders in order of decreasing length, and stored the list. This was repeated for 11 replicate cells. Finally

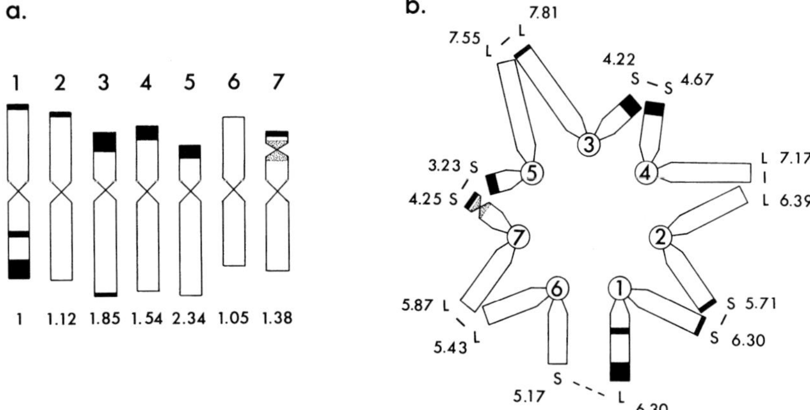

Fig. 3. The haploid karyotype of *Secale africanum* showing the major C-bands (a) arranged traditionally in order of decreasing chromosome size (from left to right); arm volume ratios are given below each chromosome; and (b) the natural karyotype ordered according to Bennett's model (numbers by arms are their mean volumes in μm^3) reproduced from Bennett (1982).

the computer summed the rank for each order from each replicate cell, reranked the 360 sums of ranks, and printed them out in increasing size order. Sums for individual routes varied widely (from 1,183 to 2,853), but the predicted order (1-2-4-3-5-6-7) ranked first out of 360 for *Secale africanum* genomes in their F1 hybrid with *Hordeum vulgare* (Bennett, 1982).

Using computer analyses similar to that already described, we showed that in pooled data for about 10 cells, the mean orders of chromosomes predicted by this model were the best expressed out of all 360 possible orders, in five out of five tests involving the haploid complements of four diploid grass species: *Aegilops umbellulata, Hordeum vulgare, H. bulbosum* and *Secale africanum* (Heslop-Harrison and Bennett, 1983). All four were tested in root tip cells, and *H. vulgare* was also tested in archesporial cells at premeiotic mitosis (Bennett, 1984b).

The chance of a predicted order ranking 1st out of 360 in 4 materials out of 4, is remote. These results establish that there is a geometric tendency for all the chromosomes in a simple haploid genome to display secondary associations and to have relatively fixed positions in relation to each other. Moreover, they demonstrate an ability of the model to predict correctly the mean spatial order of centromeres in haploid genomes of different materials. This is important because it points towards the fundamental mechanisms in nuclear architecture determining that order, which include controls of chromosome arm size, and hence of arm DNA content (Bennett, 1982). This work, including its statistical validity, was strongly criticised (Callow, 1985). However, such criticism, even if valid, misses the fundamental biological point. Exactly the same procedure was followed for all possible orders. Thus, there is no reason inherent in the method which we used to rank orders, why it should tend to place any order of centromeres (including the predicted one) first, or even high, in a ranked list of all possible orders. Yet it consistently did so, in several independent tests on different materials. The biological basis of this phenomenon needs an explanation.

As nature appears to construct karyotypes based on common principles (the naturally occurring mean spatial order of chromosome types for each basic haploid genome), Bennett (1984a) defined such an order as the 'natural karyotype'. If so, the nucleotypic basis of the model, and the existence of a natural karyotype for each basic genome, may together provide another important concept for unifying our perception and portrayal of plant genomes. Indeed, the available evidence indicated that the natural karyotype may provide the basis for unifying understanding and interrelating variation in genomic characters as diverse as the gene map, the distribution of C-bands, aspects of chromosome mechanics, karyotype architecture and genome evolution (Bennett, 1984a).

This conceptual approach led to the plan at Kew '...to address some very broad questions regarding plant genomes, including: (1) is there an ancestral angiosperm genome? If so (2) what is the ancestral angiosperm gene map? (3) Are there general genetic, mechanical, or physical principles which affect, or constrain, karyotype structure and evolution?' (Bennett, 1990).

THE ANCESTRAL GENOME

The natural karyotype may also provide an acceptable new system of karyotype nomenclature with a sound biological basis. Bennett (1984a) considered how best to represent the natural karyotype for a species, and how best to compare natural karyotypes for different related and unrelated species. Regarding the second question it was stated: 'There are good reasons to expect a tendency for related chromosome segments and DNA sequences to occur at physically similar locations within the natural karyotype...of related or even unrelated species.' Thus 'In order to facilitate biosystematic comparisons of widely unrelated taxa, it may also be meaningful to compare genomes portrayed as either a straight line or as a circle of constant relative length (100%). The latter might be most useful as it would facilitate comparisons with the circular DNA molecules of bacterial genomes. For eukaryotes the circle would be subdivided into arcs representing chromosome arms according to their relative physical sizes or DNA contents, and according to their order in the natural karyotype, with the discontinuity in a constant position. The physical locations of a feature could then be given not only within a chromosome, but also (for the first time) in terms of a relative position in the genome as a whole, independent of the number of chromosomes in the natural karyotype... Moreover, interspecific comparisons of genomes portrayed in this way would show whether or not some features of genome architecture are relatively independent of variation in characters such as genome size or basic chromosome number (x).' Fig. 4 (reproduced from Bennett, 1984a) shows the natural karyotype for *Hordeum vulgare* cv. 'Sultan' portrayed diagramatically as a circle.

Such work, theorizing about the portrayal and comparisons of natural karyotypes, led me to ask whether there might be a basic or ancestral genome structure common to all angiosperms and, if so, how to demonstrate its existence and discover its order and form. I reasoned then, that if a basic or ancestral angiosperm genome order exists, then species with only two chromosomes (or ideally only one chromosome) should reveal it most easily. For a haploid complement with only 2 chromosomes, there are only two ways of arranging the chromosomes arms in a concept circle (only one being the natural karyotype).

In 1984 only four angiosperms with 2n = 4 were known: two monocots (in the Gramineae), and two dicots (in the Compositae). Then, as now, no angiosperm with only one chromosome pair was known. It was further reasoned that if comparing circularized genetic maps for such species from these widely unrelated monocot and dicot familes showed a high overall level of conserved gene order, then this would constitute convincing evidence that a basic or ancestral genome order for angiosperms exists, and also reveal its linear form. This thinking was the basis of research on these species, describing (Bennett et al., 1986) an accurate karyotype based on EM reconstructions for the grass *Zingeria biebersteiniana* (2n = 4), and taking the first steps to construct a physical map for this species, using in situ hybridisation to localize rDNA genes (Bennett et al., 1995). This experimental approach to studying the ancestral genome, outlined at the Stebbins Symposium (AIBS, Toronto) in 1989, still seems valid (Bennett et al., 1995).

ARE CIRCULARIZED 'ANCESTRAL' CEREAL GENOMES RELATED TO NATURAL KARYOTYPES?

Moore et al. (1995a,b) recently described a hypothetical single ancestral cereal chromosome (Fig. 5), and conceptualised circular alignments of six unrelated cereal genomes which all reflect the same circularized ancestral grass genome (Fig. 6). These exciting ideas raise important questions. The first is: are the conceptualised alignments of cereal genomes, which reflect the circularized ancestral grass genome, natural karyotypes? Unfortunately it is impossible to answer this question in 5 of the 6 taxa for which such circularized genomes were published, as accurate estimates of chromosome arm sizes, essential for predicting natural karyotypes, are unavailable. However, both accurate chromosome arm sizes and conceptual circularized genomes are available for *Zea mays*, so a meaningful comparison is possible for this material.

The relative sizes of the ten chromosomes and of their arms in reconstructed root-tip metaphase cells of *Z. mays* cv. Seneca 60 were published recently (Bennett and Laurie, 1995). The data are identical with those used previously (Bennett, 1983) to try and predict the natural karyotype of maize. Bennett (1983) concluded: 'Attempts to predict the mean spatial order of the 10 haploid chromosomes of *Zea mays*, using my model, do not produce a convincing single order involving all 10 chromosomes with a single discontinuity. Instead, the model strongly suggests that there are two discontinuities and two sub-sets of 5 chromosomes'. This was contrary to the then generally accepted view that 'maize is a true diploid' as Callow (1985) critically noted.

Bennett (1984a) also suggested a 'probable mean order of chromosomes' for the two sub-sets. The Bennett model was claimed to be able to predict the natural karyotype given one complete basic set of chromosomes, and highly accurate arm sizes. It could not, however, predict the division of chromosomes between donor genomes in a polyploid, which would need many new assumptions. Thus, the original attempt to predict the full two subsets in maize (Bennett, 1984a) was invalid, as it went far beyond the rules of the model developed for diploids. However, that mistake should not divert attention from the above mentioned valid prediction of the model, since proved correct, that maize is indeed a full tetraploid.

Moore et al. (1995a) stated: 'The analysis confirms that maize is a complete tetraploid although the two genomes are rearranged relative to one another. Interestingly, most of the larger chromosomes (1, 2, 3, 4 and 6) form one genome, and a shorter set (5, 7, 8, 9 and 10) form the other.' Thus, the identities of all the chromosomes in the two genomes of maize appear to be known. If so, it is now possible to predict the natural karyotype for each *Zea mays* genome using the Bennett model. Moreover, Moore et al. (1995a) showed the sequence

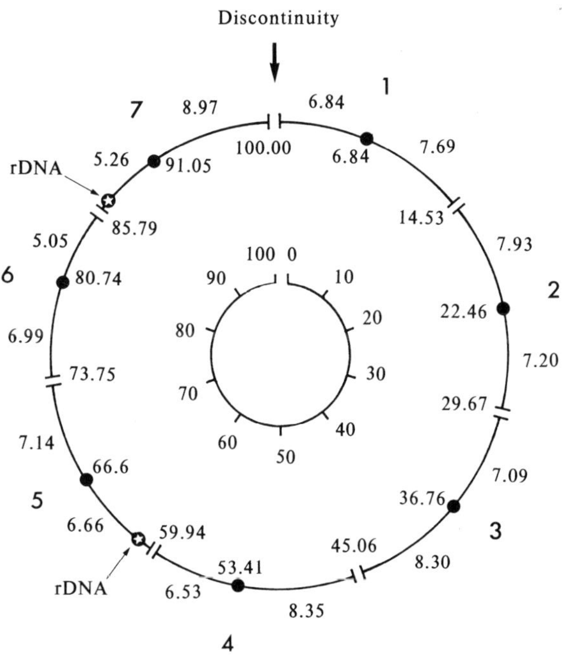

Fig. 4. The natural karyotype of *Hordeum vulgare* cv. 'Sultan' portrayed diagrammatically as a circle (circumference = 100 percentage units, with clockwise polarity) reproduced from Bennett (1984a). Chromosomes numbered 1 to 7 (large numbers outside the circle) in a clockwise direction, starting at the discontinuity, are represented as arcs. Small numbers outside the circle give the relative volumes of the 14 arms (as a percentage of the total haploid set). The positions of telomeres (radial lines at the end of chromosome arcs), centromeres (solid circles) and nucleolar organisers (starred circles) are given (numbers inside the circle) as the cumulative percentage of arm volumes moving clockwise from the discontinuity.

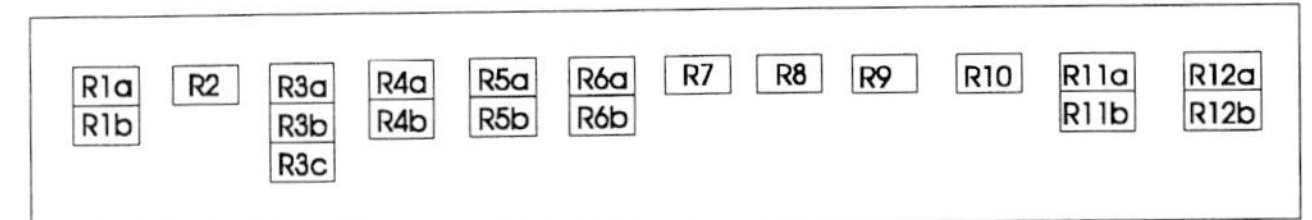

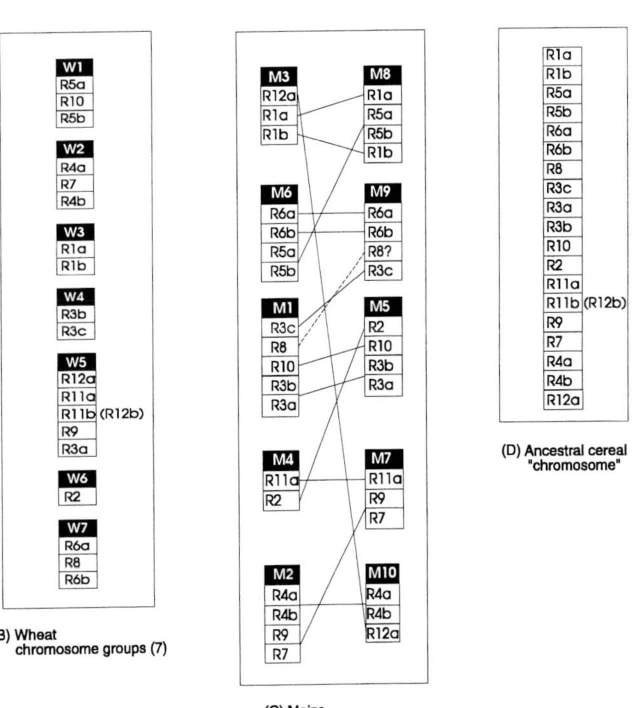

Fig. 5. Comparisons of cereal genomes based on rice linkage segments (reproduced from Fig. 1 in Moore et al., 1995a,b). (a) Rice chromosomes dissected into linkage blocks. (b) Wheat, and (c) maize chromosomes represented as 'rice blocks' on the basis of homology and/or conservation of gene order. Connecting lines indicate duplicated segments within the maize chromosomes. (d) An ancestral 'single chromosome' reconstructed on the basis of these linkage blocks.

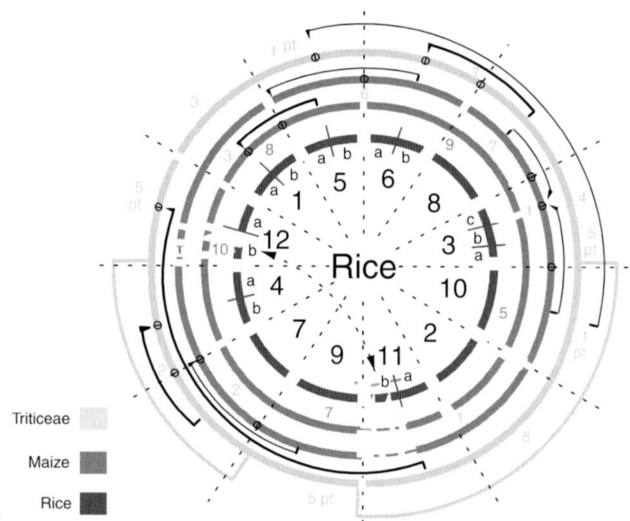

Fig. 6. Alignment of the genomes of wheat maize and rice with 19 rice linkage segments, whose order reflects the circularized ancestral grass genome (adapted from Fig. 2 in Moore et al., 1995a). The data in Fig. 5 have been redrawn as a series of rice linkage segments (defined by radiating lines) formed into chromosomes (broken numbered arcs). The thin dashed lines correspond to the duplicated segments. Linkage segments forming parts (pt) of wheat chromosome 5 are shown as a series of segments connected by dotted lines. The alignment is based on the genetic map of the D genome of wheat. The thin line indicates the duplicated segments shown as blocks 11b and 12b. Chromosomes formed by the insertion of one segment into another are shown by lines with arrows indicating the direction and point of insertion. The points of chromosome breakage involved with insertion events are indicated by black bisected circles. Numbers identify chromosomes in the haploid genomes of rice (■, centre), wheat (□, peripheral), and maize (■, two intermediates).

of the five chromosomes in each genome aligned in a conceptual circle which reflects the circularized ancestral grass genome (Fig. 6). The order in the larger genome was 1, 4, 2, 3, 6 (and back to 1 again), while in the smaller genome it was 5, 7, 10, 8, 9 (and back to 5 again). An interesting question is, therefore: how do these circularized ancestral orders relate to the natural karyotypes for these two genomes predicted using the Bennett model?

Fig. 7a shows sizes in megabases for each chromosome arm in the larger genome of maize. In Fig. 7c long and short arms are separated, and pairs of long arms and pairs of short arms are associated, as required by the Bennett model, in the unique way which minimizes the sum of differences between associated arms for the genome (omitting the discontinuity). Fig. 7e draws it as a linear array, with pairs of long arms and pairs of short arms associated in the one sequence which gives the minimum sum of differences (69). The chromosomes have the order 1, 4, 2, 3, 6 in this natural karyotype. Circularizing retains this order, and puts 1 next to 6. Comparison shows that the order reflecting the ancestral genome (1, 4, 2, 3, 6 and back to 1 again; Fig. 6), is identical with the natural karyotype for this genome predicted by the Bennett model (Fig. 7e).

Fig. 7 also presents a similar comparison for the smaller *Zea mays* genome. Fig. 7b shows the sizes in megabases for each chromosome arm in the smaller genome of maize. In Fig. 7d,

long and short arms are separated, and pairs of long arms, and pairs of short arms, are associated, according to the Bennett model. Fig. 7f shows this as a linear array with pairs of long arms and pairs of short arms associated in the one sequence which gives a minimum sum of differences (21). The chromosomes have the order 5, 7, 9, 10, 8 in this natural karyotype. Circularizing retains this order, but puts 5 next to 8. Comparison shows that the circularized ancestral order in the smaller genome (5, 7, 10, 8, 9 and back to 5 again) is not identical with the prediction of the Bennett model (5, 7, 9, 10, 8) but varies from it by only one difference involving the position of chromosome 9. As 8 out of 11 possible orders of 5 chromosomes show such a difference, this is not strong evidence for similarity, but nor does it deny it.

Fig. 8 summarizes these comparisons for the two genomes. The prediction of the Bennett model was identical with the ancestral genome order for the larger genome, while there was one difference (a transposition) between them for the smaller genome. Clearly there is some similarity. Indeed, this degree of similarity would occur by chance only 1 in 18 times (6.06%). Thus, the circularized alignment of the maize genomes to match the ancestral cereal genome order may be related to the order of centromeres in their natural karyotypes predicted by the Bennett model. While this remains an open question, further work to test this possibility is needed both for maize (to compare arm positions), and for other cereal species such as barley, rye and wheat.

Bennett (1984b,c) viewed the genome as a dynamic entity showing considerable variation in chromosome disposition and orientation centred around aspects of the natural karyotype. The Bennett model uses mainly arm lengths to predict a geometrical mean spatial arrangement of centromeres. Given this order of centromeres, dispositions of chromosomes arms at interphase could include: (1) an arrangement with most similar physical pairs near together; or (2) (by rotating alternate chromosomes by 180°) with pairs of very dissimilar (long and short) arms together at interphase (Bennett, 1982). Our tests on reconstructed metaphases purposefully addressed only the disposition of centromeres, as they alone are constrained, by spindle attachments, at this stage. Such tests support the prediction of the Bennett model for centromeres. Tests of telomere (arm) positions were not made, because arms from different chromosomes can show considerable independent movements before and during metaphase. To be meaningful, tests of arm positions must use undistorted, 3-D nuclei, at interphase. This was technically impossible, as chromosome arms could not be identified in such cells, until recently.

The associations of arms created in making the prediction are theoretical. Whether these associations of arms, or other ones allowed by the same centromere sequence, are formed, or are critical, at any stage or in any cell type in vivo, is still unknown.

As noted above, the only aspect of the Bennett model yet tested, concerns centromere dispositions. Thus, if a particular centromere sequence, which would allow various arm associations, is common to both the natural karyotype and the so-called ancestral genome order, then it will be important to discover the basis of that relationship, including whether it: (1) depends on a particular arrangement of entire arms, or of some segments only; (2) is spatially expressed in living cells; and if so (3) requires the same or different arm associations to

Large genome

(a)

Chromosome	Arm size (Mb)	
	Long (L)	Short (S)
1	184	153
4	166	105
2	155	123
3	168	91
6	137	71

(c)

			difference (d)
Long arm	1L	184	⎤ 18
	4L	166	⎦
	3L	168	⎤ 13
	2L	155	⎦
	6L	137	- discontinuity
Short arm	1S	153	- discontinuity
	2S	123	⎤ 18
	4S	105	⎦
	3S	91	⎤ 20
	6S	71	⎦

sum d = 69

(e)

Size (Mb) 153 184 166 105 123 155 168 91 71 137
d   18   18   13   20

Small genome

(b)

Chromosome	Arm size (Mb)	
	Long (L)	Short (S)
5	131	119
7	136	63
10	113	57
8	146	56
9	114	77

(d)

			difference (d)
Long arm	8L	146	- discontinuity
	7L	136	⎤ 5
	5L	131	⎦
	9L	114	⎤ 1
	10L	113	⎦
Short arm	5S	119	- discontinuity
	9S	77	⎤ 14
	7S	63	⎦
	10S	57	⎤ 1
	8S	56	⎦

sum d = 21

(f)

Size (Mb) 119 131 136 63 77 114 113 57 56 146
d   5   14   1   1

Fig. 7. Predicting the natural karyotypes for the two genomes of maize using the Bennett model. (a and b) DNA contents (in megabases) of each chromosome arm (from Bennett and Laurie, 1995) for the (a) larger and (b) smaller genomes. (c and d) Pairs of long arms, and pairs of long arms associated in the unique way which minimizes the sum of differences between associated arms for the (c) larger, and (d) smaller genomes (omitting the discontinuity). (The difference, d, between the members of each pair is shown, and their total summed, Σd.) (e and f) The resulting natural karyotypes drawn as linear arrays for the (e) larger, and (f) smaller genomes.

assemble both a circularized ancestral genome and the natural karyotype for a species.

While scientists persist in representing genomes in various 2-D ways, it must be remembered that they are all artificial, as the genome always has a 3-D form in vivo. A circularized genome reflecting an ancestral cereal genome is an interesting idea, but is its existence merely conceptual ('shape without form'), or does it have some structural reality in grass nuclei? Aspects of the natural karyotype are claimed to exist as a geometrical feature, based on analysis of mean arrangements of centromeres in reconstructed nuclei. Thus, if close relationships between natural karyotypes and the ancestral genome occur in other species besides maize, this would infer that the ancestral genome also has some 3-D form in their nuclei.

A second important question posed by the work of Moore et al. (1995a,b) is: Why do genomes, thought to have diverged over 60 million years ago, still all retain in their structure such a strong overall expression of a common genome order? Moore et al. (1995b) suggested that it reflects a common ancestral

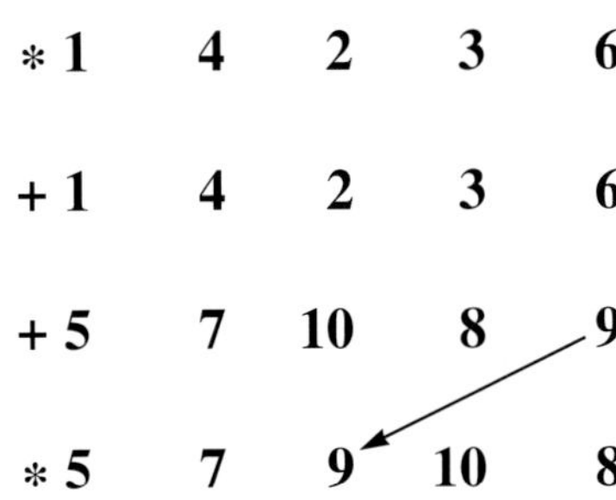

Fig. 8. Natural karyotypes for the two genomes of maize predicted using the Bennett model, compared with their 'ancestral' orders shown as circularized genomes in Fig. 5 (from Moore et al., 1995a), but drawn here as a linear array.

genome, perhaps comprising a single chromosome. Given the many processes known, such as inversions and translocations, which generate change and tend to disrupt any primeval genome order, it seems unlikely that so much detectable expression of any ancestral order would remain after so long without some strong selection against disruptive change. If selection is responsible for maintaining the order, on what would it act?

An alternative rationale for a common order emphasises common function more than common ancestry. Recent research increasingly suggests that higher order genome disposition can be important for gene expression (Bennett, 1984c; Cremer et al., 1993; Heslop-Harrison et al., 1993). Perhaps the reason why it is possible to assemble the linkage segments from species with widely diverged chromosome numbers and genome sizes into a common order, is not primarily a reflection of their common ancestry, but because this arrangement reflects some common functional 3-D form(s) of the angiosperm genome assembled in cells, at one or more stages, in all the species. Selective forces acting at the level of the genome, could be a fundamentally similar internal homoeostasis that recognises and selects for a particular type of genome organisation. Indeed, selection for this ability may operate, either because it is essential, or just the fittest solution (so far), for their survival. If so, knowledge of this structure would significantly unify our understanding of genomes and their evolution, and may explain phenomena as diverse as karyotype orthoselection, spatial separation of haploid genomes, the natural karyotype and the ancestral genome.

REFERENCES

Bennett, M. D. (1971). The duration of meiosis. *Proc. Roy. Soc. Lond. B* **178**, 277-299.

Bennett, M. D. (1972). Nuclear DNA content and minimum generation time in herbaceous plants. *Proc. Roy. Soc. Lond. B* **181**, 109-135.

Bennett, M. D. (1973). Nuclear characters in plants. *Symp. Biol.* **25**, 344-366.

Bennett, M. D. (1976). DNA amount, latitude and crop plant distribution. *Env. Exp. Bot.* **16**, 93-108.

Bennett, M. D. and Smith, J. B. (1976). Nuclear DNA amounts in angiosperms. *Phil. Trans. Roy. Soc. Lond. B.* **274**, 227-274.

Bennett, M. D. (1977). The time and duration of meiosis. *Phil. Trans. Roy. Soc. Lond. B* **277**, 201-226.

Bennett, M. D., Smith, J. B., Ward, J. and Jenkins, G. (1981). The relationship between nuclear DNA content and centromere volume in higher plants. *J. Cell Sci.* **47**, 91-115.

Bennett, M. D. (1982). Nucleotypic basis of the spatial ordering of chromosomes in eukaryotes and the implications of the order for genome evolution and phenotypic variation. In *Genome Evolution* (ed. G. A. Dover and R. B. Flavell), pp. 239-261. Academic Press, London.

Bennett, M. D. (1983). The spatial distribution of chromosomes. In *Kew Chromosome Conference II* (ed. P. E. Brandham and M. D. Bennett), pp. 71-79. Allen & Unwin, London.

Bennett, M. D., Heslop-Harrison, J. S., Smith, J. B. and Ward, J. P. (1983). DNA density in mitotic and meiotic metaphase chromosomes of plants and animals. *J. Cell Sci.* **63**, 173-179.

Bennett, M. D. (1984a). The genome, the natural karyotype, and biosystematics. In *Plant Biosystematics* (ed. W. F. Grant), pp. 41-66. Proc. Symp. Int. Org. Plant Biosyst. July 1983, McGill University, Montreal, Canada. Academic Press, Canada.

Bennett, M. D. (1984b). Premeiotic events and meiotic chromosome pairing. *SEB Symp.* **38**, 87-121.

Bennett, M. D. (1984c). Nuclear architecture and its manipulation. In *Gene Manipulation in Plant Improvement* (ed. J. P. Gustafson), pp. 469-502. 16th Stadler Genetics Symp. Plenum Publishing Corporation, New York.

Bennett, M. D. (1985). Intraspecific variation in DNA amount and the nucleotypic dimension in plant genetics. In *Plant Genetics* (ed. M. Freeling), pp. 283-302. UCLA Symp. Mol. Cell. Biol. New Series, vol. **35**. New York, Alan Liss.

Bennett, M. D., Smith, J. B. and Seal, A. G. (1986). The karyotype of the grass *Zingeria biebersteiniana* (2n=4) by light and electron microscopy. *Can. J. Genet. Cytol.* **28**, 554-562.

Bennett, M. D. (1987). Variation in genomic form in plants and its ecological implications. *New Phytologist* **106** (suppl.), 177-200.

Bennett, M. D. (1988). Parental genome separation in F1 hybrids between grass species. In *Proceedings of the Third Kew Chromosome Conference* (ed. P. Brandham), pp. 195-208. HMSO, London, UK.

Bennett, M. D. (1990). Scientific visiting group. The Jodrell Laboratory. p. 59.

Bennett, M. D. and Laurie, D. A. (1995). Comparison of maize and sorghum chromosome size using serial thin-section reconstructions of root-tip metaphase nuclei. *Maydica* **40**, 199-204.

Bennett, S. T. and Leitch I. J. (1995). Nuclear DNA amounts in angiosperms. *Ann. Bot.* **76**, 113-176.

Bennett, S. T., Leitch I. J. and Bennett, M. D. (1995). Chromosome identification and mapping in the grass *Zingeria biebersteiniana* (2n = 4) using fluorochromes. *Chrom. Res.* **3**, 101-108.

Callow, R. S. (1985). Comments on Bennett's model of somatic chromosome disposition. *Heredity* **54**, 171-177.

Cremer, T., Kurz, A., Zirbel, R., Dietzel, S., Rinke, B., Schrock, E., Speicher, M. R., Mathieu, U., Jauch, A., Emmerich, P., Scherthan, H., Ried, T., Cremer, C. and Lichter, P. (1993). Role of chromosome territories in the functional compartmentalization of the cell nucleus. *CSH Symp. Quant. Biol.* **58**, 777-791.

Finch, R. A., Smith, J. B. and Bennett, M. D. (1981). Hordeum and Secale mitotic genomes lie apart in a hybrid. *J. Cell Sci.* **52**, 391-403.

Heslop-Harrison, J. S. and Bennett, M. D. (1983). Prediction and analysis of spatial order in haploid chromosome complements. *Proc. Roy. Soc. Lond. B* **218**, 211-223.

Heslop-Harrison, J. S., Leitch, A. R. and Schwarzacher, T. (1993). The physical organisation of interphase nuclei. In *The Chromosome* (ed. J. S. Heslop-Harrison and R. B. Flavell), pp. 221-232. Bios, Oxford.

Kenton, A., Dickie, J. B., Langton, J. B. and Bennett, M. D. (1990). Nuclear DNA amount and karyotype symmetry in *Cypella* and *Hesperoxiphion* Tigridieae; Iridaceae). *Evolutionary Trends in Plants* **4**, 59-69.

Leitch, A. R., Schwarzacher, T., Mosgoller, W., Bennett, M. D. and Heslop-Harrison, J. S. (1991). Parental genomes are separated throughout the cell cycle in a plant hybrid. *Chromosoma* **101**, 206-213.

Moore, G., Devos, K. M., Wang, Z. and Gale, M. D. (1995a). Grasses, line up and form a circle. *Curr. Biol.* **5**, 737-739.

Moore, G., Foote, T., Helentjaris, T., Devos, K., Kuruta, N. and Gale, M. D. (1995b). Was there a single ancestral cereal chromosome? *Trends Genet.* **11**, 81-82.

Rees, H. (1984). Nuclear DNA variation and the homology of chromosomes. In *Plant Biosystematics* (ed. W. F. Grant), pp. 87-96. Proc. Symp. Int. Org. Plant Biosyst. July 1983, McGill University, Montreal, Canada. Academic Press, Canada.

Schwarzacher-Robinson, T., Finch, R. A., Smith, J. B. and Bennett, M. D. (1987). Genotypic control of centromere positions of parental genomes in *Hordeum × Secale* hybrid metaphases. *J. Cell Sci.* **87**, 291-304.

Schwarzacher, T., Heslop-Harrison, J. S., Anamthawat-Jonsson, K., Finch, R. A. and Bennett, M. D. (1992). Parental genome separation in reconstructions of somatic and premeiotic metaphases of *Hordeum vulgare × H. bulbosum. J. Cell Sci.* **101**, 13-24.

Van't Hof, J. and Sparrow, A. H. (1963). A relationship between DNA content, nuclear volume, and minimum mitotic cycle time. *Proc. Nat. Acad. Sci. USA* **49**, 897-902.

Printed in Great Britain © The Society for Experimental Biology 1996
SEB0029

Genetic recombinational and physical linkage analyses on slash pine

R. L. Doudrick

Southern Institute of Forest Genetics, USDA Forest Service, Saucier, Mississippi, 39574-9344, USA

SUMMARY

Slash pine is native to the southeastern USA, but is commercially valuable world-wide as a timber-, fiber- and resin-producing species. Breeding objectives emphasize selection for fusiform rust disease resistance. Identification of markers linked to genetic factors conditioning specificity should expand our knowledge of disease development. Towards this end, random amplified polymorphic DNA (RAPD) markers were identified and mapped in a tree hypothesized to be homozygous dominant for resistance at one locus and homozygous recessive at another. Because the DNA prepared for analysis was from haploid maternally-inherited, megagametophyte tissue of seeds, RAPD markers were observed as either present or absent. The analysis revealed 13 linkage groups of three or more loci, ranging in size from 28 to 68 cM, and nine linked pairs. The 22 groups and pairs included 73 RAPD markers and covered a genetic map distance of ~782 cM. Genome size estimates, based on linkage data, range from 2,880 to 3,360 cM, and equal $6.0\text{-}6.9\times10^6$ bp/cM (physical size >20,000 Mbp). Using a 30 cM map scale and including unlinked markers, ends of linkage groups, and linked pairs, the RAPD markers account for ~2,160 cM or 64-75% of the genome. Mapping 80 additional RAPD markers placed 131 loci total in 20 linkage groups of three or more loci, nearly doubling the coverage in the groups to a genetic map distance of ~1,347 cM. Two other slash pine trees also have been RAPD mapped.

DNA-DNA in situ hybridization and cytochemical staining are being used to integrate the genetic recombinational maps. A karyotype and ideogram have been prepared for slash pine ($2n=2x=24$); metaphase chromosome preparations show 11 pairs of long metacentric chromosomes and one shorter pair of submetacentric chromosomes. Patterns of fluorescence in situ hybridization to genes for the large and small rRNA subunits and fluorochrome banding patterns using the GC-base-specific chromomycin A_3 (CMA) and AT-base-specific 4',6-diamidino-2-phenylindole (DAPI) allowed all twelve pairs of chromosomes to be identified and a standard karyotype established.

A family of sequences associated with $(TTTAGGG)_n$ related repeats has been identified in slash pine using a labeled synthetic oligonucleotide probe. Fluorescence in situ hybridization shows a weak signal at telomeres and significantly stronger intensity at non-telomeric sites. The most common non-telomeric location was in the pericentric regions of chromosomes; interstitial sites of hybridization were relatively common.

Microsatellite DNAs, an abundant retrotransposon-like element, and total genomic in situ hybridization and species and chromosome specific DNAs are being evaluated for analyses of interspecific hybrids and chromosome evolution between related species. Interest in low and single copy sequences is increasing.

Key words: Genetic mapping, *Pinus elliottii*, Telomere, In situ hybridization, Karyotype, rDNA

INTRODUCTION

Slash pine (*Pinus elliottii* Engelm. var. *elliottii*) is native to the southeastern United States. It is a major forest component on flatwoods and along streams and the edges of swamps and bays, wherever soil moisture is ample (Lohrey and Kossuth, 1990). Slash pine is planted commonly inside and outside its natural range (Boyer and South, 1984), as well as in exotic plantings in Asia (Pan, 1989), South America (Picchi and Barrett, 1967), Africa and Australia (Mullin et al., 1978). It is commercially valuable world-wide as a timber-, fiber- and resin-producing species.

In the United States, fusiform rust, caused by *Cronartium quercuum* (Berk.) Miyabe in Shirai f. sp. *fusiforme*, continues to be the most damaging disease on slash pine (USDA Forest Service, 1994). Control measures focus primarily on improving planting stock through selection of open-pollinated families that exhibit low percentage infections using composites of spore collections of *C. q. fusiforme* (Zobel et al., 1971). Unfortunately, this approach provides little genetic information. Artificial inoculation studies suggest specific genetic factors in host and pathogen contribute to disease development (Snow et al., 1975; Griggs and Walkinshaw, 1982; Kinloch and Walkinshaw, 1991; Doudrick and Nelson, 1993; Nelson et al., 1993a). Identification of markers linked to factors conditioning specific resistance in slash pine should expand our knowledge of the micro-coevolution of host and pathogen and assist breeders (Nance et al., 1992).

GENETIC RECOMBINATIONAL LINKAGE ANALYSES

Towards this end a molecular genetic recombinational linkage map of the nuclear genome in a mature slash pine tree was constructed using random amplified polymorphic DNA (RAPD) markers (Nelson et al., 1993b). The tree was hypothesized to possess at least four resistance loci; apparently homozygous dominant for resistance at one locus, homozygous recessive at a second locus and heterozygous at two additional loci (Doudrick and Nelson, 1993; Nelson et al., 1993a). The total number and allellism of resistance loci in slash pine are unknown, and dependent on identifying additional pathotypes of *C. q. fusiforme* capable of conditioning differential interactions.

The map was constructed using a source of segregating haploid tissue unique to conifers, the female megagametophyte. This megagametophyte is nutritive endosperm resulting from mitotic divisions of a single meiotic product; the product giving rise also to the embryo of the same seed. Since the DNA prepared for analysis from the megagametophyte is haploid and strictly maternally inherited, segregation and recombination can be evaluated in a sample of wind-pollinated seeds from a single tree (Guries et al., 1978; Conkle, 1981) and the RAPD markers observed as either present or absent (Tulsieram et al., 1992).

RAPD reaction protocols were described by Welsh and McClelland (1990), and Williams et al. (1990) and modified for slash pine by Nelson et al. (1993b). To identify clear and repeatable polymorphisms, arbitrary (sequence 50-80% C+G content), 10 base, oligonucleotide RAPD primers were first screened for amplification of segments within eight megagametophyte DNA samples. These RAPDs were characterized further within a sample of 68 megagametophytes from the same tree. Segments segregating in a Mendelian ratio of 1:1 (α=0.05) by chi-square analysis (χ^2), present-to-absent, were classified and mapped using multi-point linkage analysis (MAPMAKER II version 1.9; Lander et al., 1987).

The analysis revealed 13 linkage groups of three or more loci, ranging in size from 28 to 68 recombinational units (centiMorgans, cM), and nine linked pairs of loci. The 22 groups and pairs included 73 RAPD markers that covered a genetic map distance of ~782 cM; genome size estimates ranged from ~2,880 to 3,360 cM. With a physical size of >20,000 Mbp (Dhillon, 1980; Ohri and Khoshoo, 1986; Greilhuber, 1988; Wakamiya et al., 1993) and this linkage data, values of 6.0-6.9×10⁶ bp/cM are obtained for comparative and map-based analyses. Using a 30 cM map scale and including the 24 unlinked markers and the ends of the 13 linkage groups and nine linked pairs, these RAPD markers account for ~2,160 cM or 64-75% of the genome. Because of this low percentage coverage, 80 additional RAPD markers were mapped (C. D. Nelson, personal communication), permitting placement of 131 loci total into 20 linkage groups of three or more loci, nearly doubling the coverage in the groups to a genetic map distance of ~1,347 cM (Fig. 1).

Data are preliminary on a map of another single mature slash pine; this tree is hypothesized to have a different complement of specific resistance loci. Analyses are incomplete with only 32 megagametophyte DNAs scored, but show 151 polymorphic segments segregating 1:1 (T. L. Kubisiak, personal communication).

In similar fashion, a single mature slash pine tree selected for high wood specific gravity and its progeny screened for heritability of fusiform rust disease reaction was RAPD mapped (Byram et al., 1992). The RAPD primers selected for

mapping produced 185 scoreable markers with segregation ratio 1:1. The 185 loci were placed into 19 linkage groups of three or more loci and four pairs. The resulting map covered ~1,776 cM, about 59% of the genome.

RAPDs were used also to simultaneously construct linkage maps of the mature slash pine and longleaf pine (*Pinus palustris* Mill.) parent trees of an interspecific F₁ family (Kubisiak et al., 1995). RAPD primers were screened against the two parents and six F₁ progenies. A total of 247 segregating loci (233 at 1:1; 14 at 3:1) and 87 polymorphic (between-parents), but non-segregating, loci were identified; 101 loci segregating 1:1 for the slash pine parent. The 101 marker loci were placed in 13 groups and six pairs for ~952.9 cM using multi-point linkage analysis (MAPMAKER/EXP version 3.0; Lincoln et al., 1992). When a weighted average distance between groups and unlinked markers was added with the mapped markers in the groups and pairs, genome coverage was estimated at ~1,462.9 cM. For a 30 cM map scale, this is approximately 61.7% of the total genome length, the estimated length somewhat smaller for slash pine by this method than from the maps above, approximately 2,300 cM compared to about 3,000 cM. Although Moran et al. (1983) reported sexually related regional variations in meiotic recombination in *Pinus radiata* D. Don, greater recombination rates for female vs male meiocytes is common in plants (Rhoades, 1941) and may explain the above differences. Linkage of the 3:1 loci to 1:1 loci in each of the parental maps was used to infer further linkages within maps, as well as potentially homologous counterparts between maps. Three of the longleaf-pine linkage groups appear to have potentially homologous counterparts to four different slash-pine linkage groups identified by Nelson et al. (1993b).

All the current maps of mature slash pine trees are of medium density and incomplete (more groups than chromosomes). The extent of genomic coverage in the maps does not allow for efficient mapping of genes conditioning specific resistance for *C. q. fusiforme* in this species. Obviously more markers are required to bridge gaps between current groupings and expand coverage towards 100%. Estimates of the number of markers required to obtain 90% coverage of the pine genome (average spacing of 20-30 cM) suggest that approximately 200 to 300 additional markers will be required (Nelson et al., 1993b). Although automated preparation of RAPD reactions showed improvements in speed (Nance and Schumate, 1992) and repeatability, at the current levels of genomic coverage (60-85%), screening more primers for additional polymorphisms may not be cost effective. The very high bp/cM average ratio in slash pine will require very high resolution mapping <0.1 cM average spacing for map-based cloning applications (Nelson et al., 1994). Increasing sample size and applying bulked segregant analysis (Michelmore et al., 1991) should help identify polymorphisms near termini of the groups and remaining unlinked markers.

Sampling other regions of the genome to supplement RAPD mapping efforts may be best served by integrating other types of markers (e.g. isozymes, proteins, restriction fragment length polymorphisms (RFLPs), sequence characterized sites), markers commonly used to analyze quantitative genetic variation and study genome structure and evolution. The additional markers will be especially useful to evaluate synteny with other species. Conkle (1981) using isozymes identified allelic

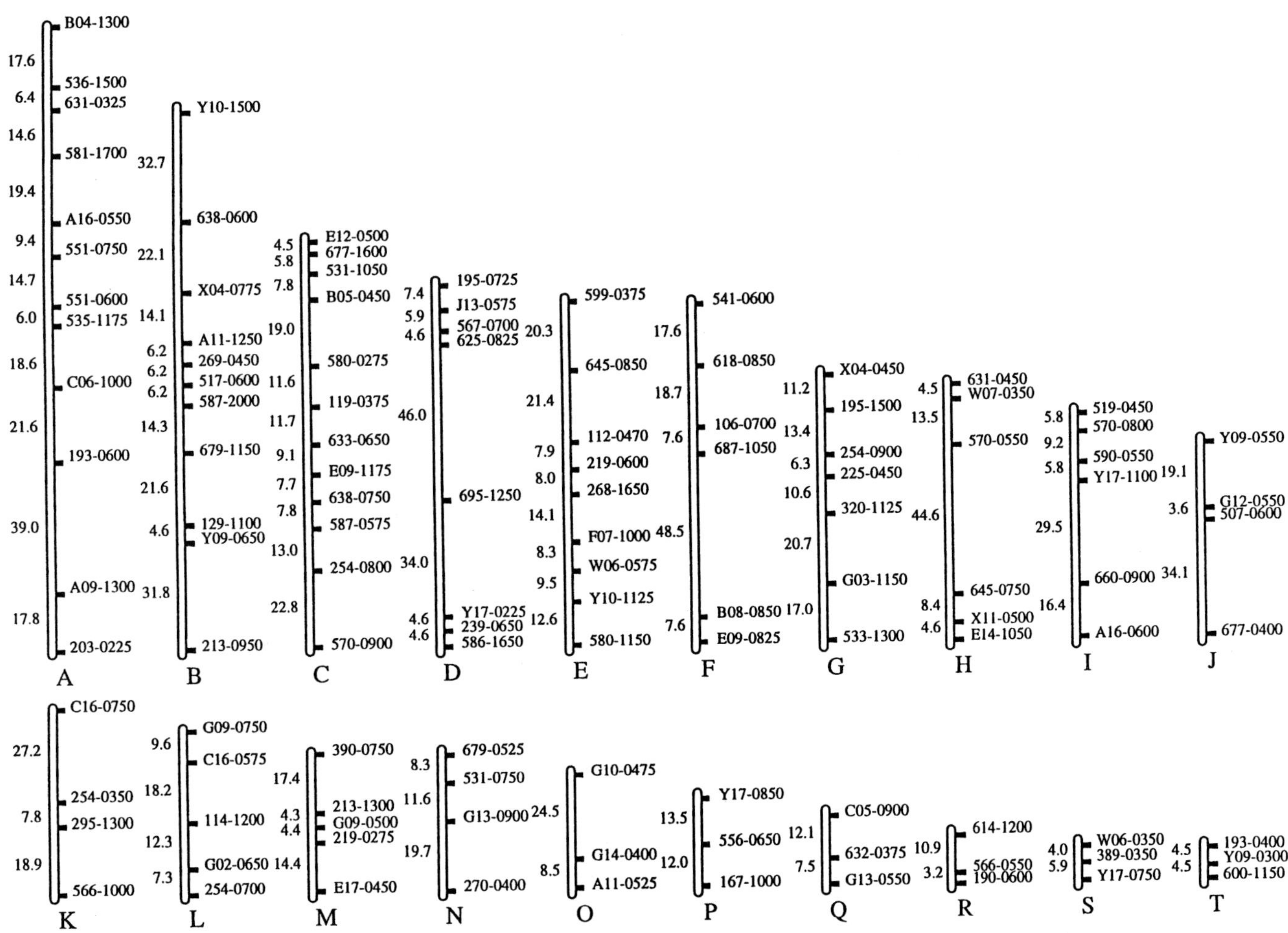

Fig. 1. RAPD genetic recombinational linkage map of a mature slash pine tree (Southern Institute of Forest Genetics clone 8-7). One thousand three hundred map units (cM) are present in linkage groups with three or more loci, indicated by letters A through T. Marker names are given on the right-hand side of the linkage groups and Haldane (Haldane function; Haldane 1919) cM distances are given on the left-hand side. The marker names contain the primer ID and approximate length in base pairs of the RAPD product. Linked loci for which a position could not be determined (at lod >1.7) are not included.

variation and linkage in five *Pinus* spp. The loci consistently mapped in the same order and with similar map distances among the species from different subsections of the genus. Similar data on slash pine would be immediately informative. Unique to slash pine are the studies reported by Michelozzi et al. (1990). They showed simple inheritance in 85 slash pine clones for high β-phellandrene, a cortical monoterpene, and demonstrated the gene was linked with factors directly involved with relatively low fusiform rust disease reaction. Both isozyme and protein phenotypes identify coding regions of the genome. Forty RFLP marker loci have been placed by J. P. van Buijtenen (unpublished data) on the mature slash pine tree map reported by Byram et al. (1992) and summarized above. These RFLPs provide another opportunity to compare the genomes of related pine species through hybridization patterns and possibly construct RFLP maps (Ahuja et al., 1994).

MOLECULAR CYTOGENETIC ANALYSES

A detailed karyotype could be used to efficiently assign recombinational map linkage groups to physical chromosomes, integrate the various maps, and complement studies on the molecular and quantitative genetics of slash pine. The lack, however, of morphological variability in chromosomes historically has impeded progress in karyotyping species of *Pinus* (Pederick, 1967). Most *Pinus* species, including slash pine (as *P. elliottii*, Mergen, 1958; as *Pinus caribaea* Morelet, Mehra and Khoshoo, 1956; Ogoshi and Nakata, 1955), have 12 pairs of chromosomes and are characterized by 10 or 11 pairs of long, metacentric chromosomes and one or two pairs of shorter submetacentric chromosomes. DNA-DNA in situ hybridization is the key to developing high-resolution techniques that can be applied reproducibly to samples, and linking molecular results with chromosomes (Heslop-Harrison et al., 1991).

Doudrick et al. (1995) have prepared a karyotype and ideogram for slash pine ($2n=2x=24$) using mitotic metaphase cells from root-tips of seedlings; each metaphase had 11 pairs of long metacentric chromosomes and one shorter pair of submetacentric chromosomes. Patterns of fluorescence in situ hybridization using the coding sequences of the genes for 18S-

5.8S-25S rRNA (18S-25S rDNA) and the complete gene for 5S rRNA and accompanying application of fluorochrome banding allowed identification of all twelve pairs of chromosomes. Fluorescence in situ hybridization using biotin-labeled pTa71 (from *Triticum aestivum* L.; Gerlach and Bedbrook, 1979) showed seven pairs of chromosomes with intercalary sites (six major sites and one of lesser intensity and therefore possibly fewer copies) for 18S-25S rDNA and one pair with a paracentromeric site (Fig. 2). The pXV1 (from *Beta vulgaris* L.; Schmidt et al., 1994) digoxigenin-labeled probe localized a major 5S rDNA hybridization site on another pair of chromosomes (Fig. 3). Two minor sites occurred on other chromosomes, one with and one without a site of 18S-25S rDNA. In both cases, a heterologous probe of DNA from an angiosperm was used as probe, thus only the sites where the rRNA coding sequence is present were identified.

After in situ hybridization, chromosomes were stained with the AT-base-specific fluorochrome 4′,6-diamidino-2-phenylindole (DAPI) and the GC-base-specific fluorochrome chromomycin A$_3$ (CMA) (Figs 2, 3). Positive CMA bands at intercalary sites were detected only on two pairs of chromosomes, paracentromeric sites only on three pairs of chromosomes, and bands at both regions on four pairs. Many bands of CMA showed at sites of hybridization of 18S-25S rDNA but one major band occurred at a centromere without an rDNA site. After staining with DAPI, bands appeared at AT-rich intercalary and/or centromeric regions of nearly all chromosomes. A characteristic double band of DAPI occurred near the centromere of one pair of chromosomes.

Patterns of fluorescence in situ hybridization and fluorochrome banding allowed all twelve pairs of chromosomes in slash pine to be identified (Fig. 4; Table 1). As reported for other species of pines (Sax and Sax, 1933), previously only a single chromosome pair, the shortest and only nonmetacentric chromosomes, could be identified unequivocally by size and centromere index alone. All 12 chromosome pairs were identified consistently after probing for the 18S-25S and 5S rDNA loci, and fluorochrome staining. The staining results using CMA and DAPI compare favorably with banding patterns

reported for *P. banksiana* Lamb. and *P. nigra* Arnold (MacPherson and Filion, 1981), *P. densiflora* Sieb. et Zucc.

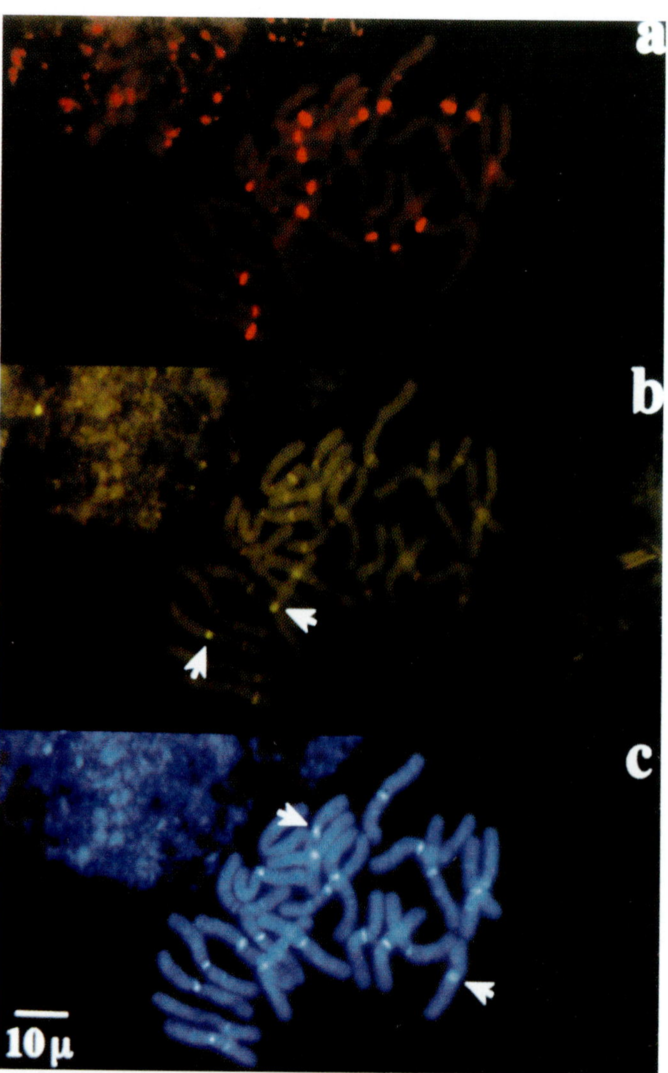

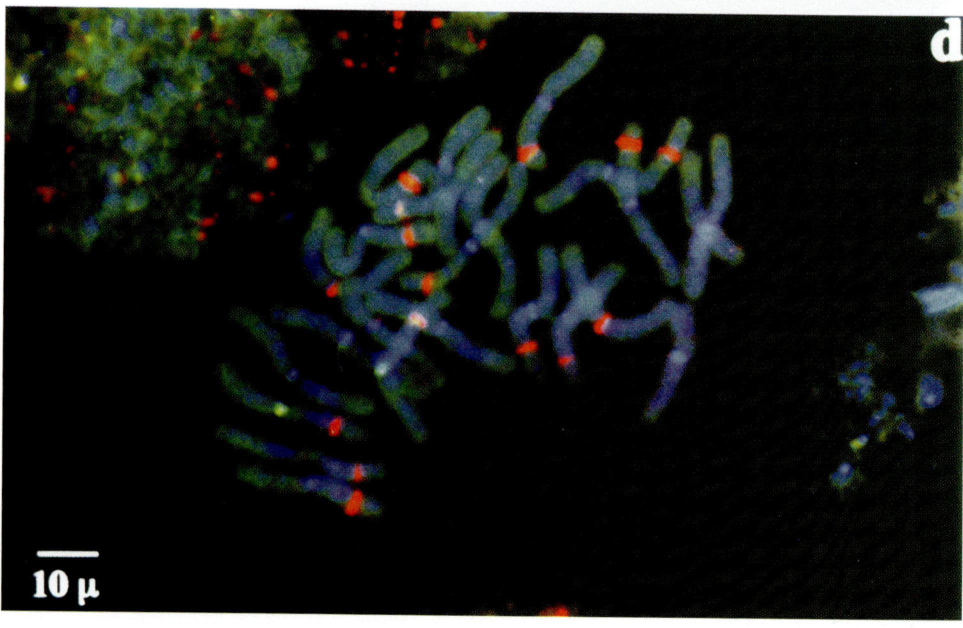

Fig. 2. In situ hybridization of rDNA and fluorochrome banding to mitotic metaphase chromosomes of root-tips of seedlings of slash pine: (a) in situ hybridization showing the eight pairs of sites of 18S-25S rDNA detected red. (b) CMA staining showing GC-enriched chromosome segments (bright yellow). Most sites are co-incident with sites of 18S-25S rDNA, but a centromeric pair on chromosome 4 is prominent but not rDNA-associated (arrows). (c) DAPI staining showing the 24 chromosomes and bright, AT-enriched bands at most centromeres. Chromosome 10 shows double bands on either side of the centromere (arrows). (d) Overlay of images a, b and c. Note also banding and hybridization sites in the adjacent interphase nucleus.

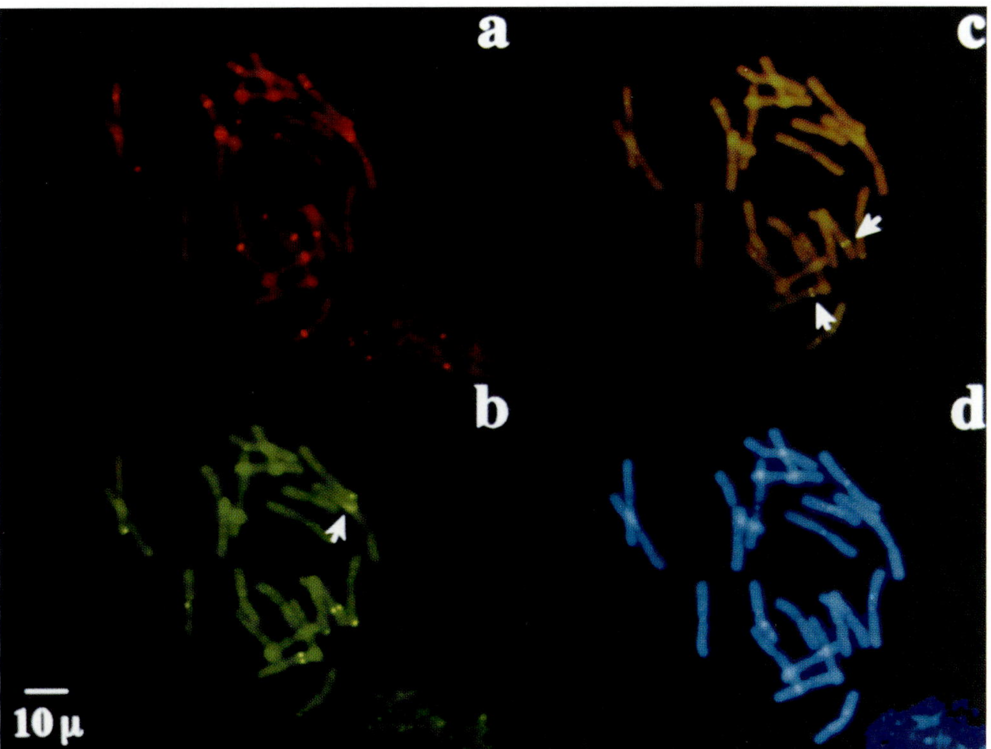

Fig. 3. In situ hybridization of rDNA and fluorochrome banding to spreads of metaphase chromosomes of root-tips of seedlings of slash pine: (a) in situ hybridization showing the eight pairs of sites of 18S-25S rDNA detected red; (b) CMA staining showing GC-enriched chromosome segments (bright yellow); (c) in situ hybridization using a probe of 5S rDNA, detected yellow-green, showing the major (chromosome 9) and two minor pairs of 5S sites (chromosomes 1 and 2; one chromosome 1 site, arrowhead, is obscured underneath another chromosome); (d) DAPI staining showing the morphology of the 24 chromosomes.

(Hizume et al., 1989), and *P. luchuensis* Mayr (Hizume et al., 1990), but are dissimilar for *P thunbergii* Parl. (Hizume et al., 1989), and *P. koraiensis* Sieb. et Zucc., *P. resinosa* Ait., and *P. strobus* L. (MacPherson and Filion, 1981). The slash pine karyotype is proposed therefore to be the standard karyotype. The numbers assigned to the chromosomes for homologous group designations should serve as the framework for homoeologous group designations of chromosomes in related *Pinus* spp. and possibly for the Pinaceae and other genera of conifers.

Telomeres are the extreme termini of chromosomes and contain tandemly repeated copies of the sequence $(T/A)_{1-4}G_{1-8}$ in all eukaryotes studied (Zakian, 1989; Blackburn, 1991). Characterizing the slash pine telomeres would further contribute to unifying the genetic recombinational linkage maps and provide anchor points for the ends of linkage groups. In situ hybridization studies showed that sequences homologous to the *Arabidopsis thaliana* (L.) Heyn. telomere repeat $(TTTAGGG)_n$ are located in the terminal regions of other plant chromosomes (Schwarzacher and Heslop-Harrison, 1991). In situ hybridization studies showed also that on a number of different plant chromosomes the TTTAGGG repeat is associated with repetitive sequences that do not appear to be localized in the terminal regions (Cheung et al., 1994).

A family of sequences related to the telomeric repeats of *A. thaliana* has been identified in slash pine. The probe was a heterogeneous population of biotinylated molecules containing repeat arrays of various lengths produced by concatenation of the simple monomers $(TTTAGGG)_5$ and $(CCCTAAA)_5$ using PCR without a template (Ijdo et al., 1991; Cox et al., 1993). Although this family of sequences contains motifs homologous to the repeats of *A. thaliana* telomeres, in situ hybridization shows extremely weak signal at telomeres (identified only by using computer enhanced quantitative pixel analysis; BDS-Image version 1.5, Oncor, MD), and significantly stronger intensity at non-telomeric sites (Fig. 5). The most common non-telomeric sites were pericentric and interstitial regions of chromosomes. Variation in signal intensity was interpreted as differences in copy number at sites of hybridization

Table 1. Criteria to identify metaphase chromosomes in spreads from root-tips of seedlings of slash pine

Chromosome	Important identification criteria
1*	Minor, terminal 5S rDNA site at end of long arm.
2*	18S-25S rDNA site on short arm. Minor terminal 5S rDNA site on the long arm.
3	18S-25S rDNA site on short arm. One of the largest chromosomes.
4*	18S-25S rDNA site on long arm. Unique, strong CMA band at centromere.
5*	18S-25S rDNA site on short arm. Unique in not having DAPI band at centromere.
6	Weak 18S-25S rDNA on short arm, weak CMA band at centromere. Not one of largest chromosomes.
7*	Unique 18S-25S rDNA site at centromere.
8	18S-25S rDNA site on short arm. Not one of largest chromosomes.
9*	Unique major 5S rDNA site at intercalary position.
10*	Unique major DAPI bands each side of the centromere.
11	18S-25S rDNA site on short arm. Small chromosome.
12*	Smallest chromosome, unique in having unequal arms.

*Chromosomes that normally can be identified by inspection and without reference to other chromosomes in the karyotype. Only invariable and major bands are described. All chromosomes, except 12, are metacentric so arms with markers may reverse (see Fig. 4).

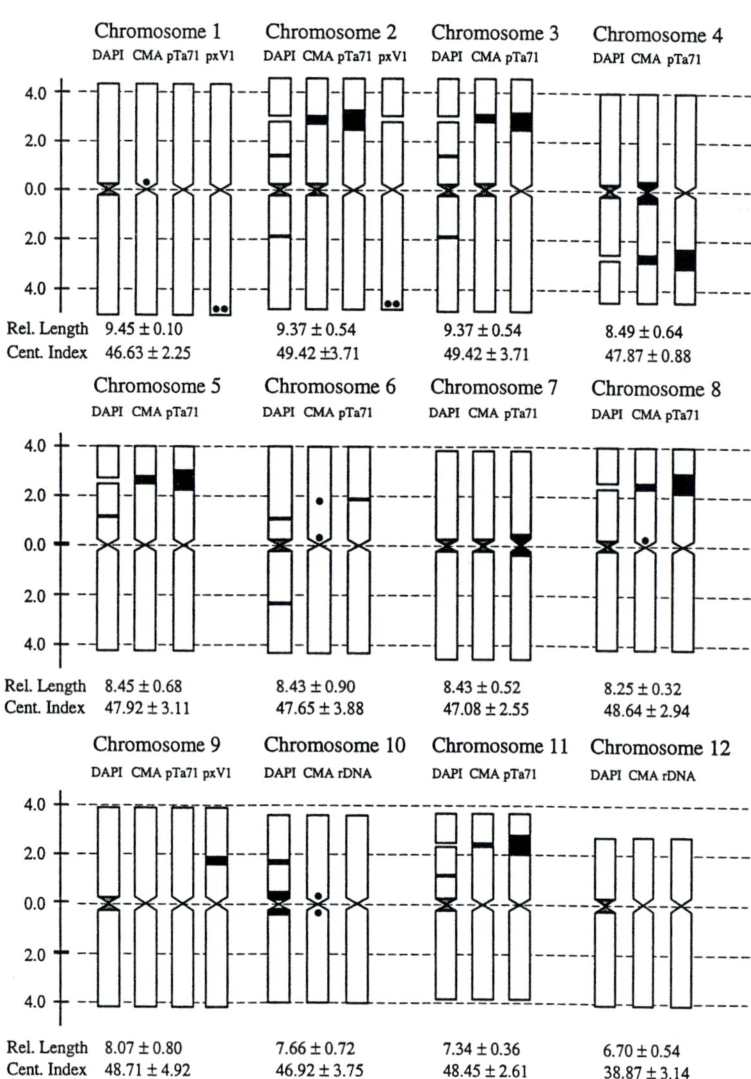

Fig. 4. Ideogram of slash pine. Darkest sites and bands show most intense fluorescence, shading less fluorescence and dots least. Band width suggests relative size of site or band. Relative lengths are percentages of the length of the total haploid chromosome complement.

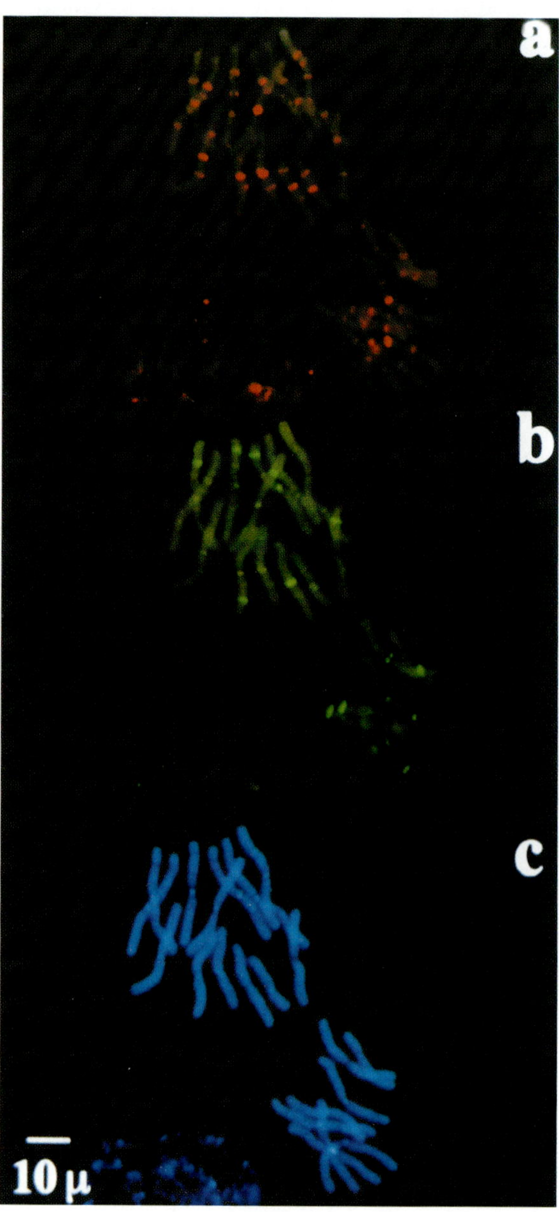

Fig. 5a-c

(Schwarzacher and Heslop-Harrison, 1991). Relatively large amounts of the sequence were present within regions of constitutive heterochromatin.

CONCLUSIONS AND FUTURE DIRECTION

The ecologic and economic importance of slash pine have prompted the construction of genetic recombinational linkage maps based principally on RAPDs (Byram et al., 1992; Nelson et al., 1993b; Kubisiak et al., 1995). Genetic mapping efforts summarized, are beginning to reveal details about the organization of coding and adjacent regions and the level of gene duplication within the genomes, thus providing a framework for comparative gene mapping.

Although such genetic maps are informative, they furnish an incomplete picture of genome organization. Molecular characterization and fluorescence in situ hybridization of non-coding sequence probes are providing some information on genomic structure. Synthetic oligonucleotides of tandem arrays of simple sequence repeats were used to detect variation in slash pine and identified polymorphic regions of the genome that distinguished slash pine from related species (A. Kamm, personal communication). A Ty1-copia retroelement was shown to be highly amplified in slash pine and widely dispersed throughout the genome (A. Kamm, personal communication). Considering the large size of the genome of slash pine and the undoubtedly high proportion of noncoding DNA, genetic mapping tasks would be very difficult without the guidance afforded by physical mapping. Physical mapping can help in the generation and use of high-resolution genetic maps by permitting: (i) the mapping of regions experiencing little or no recombination; (ii) comparison of the degree of similarity (and thus application of) genetic maps between different genotypes; (iii) assessment of multilocus probes for use in RFLP mapping; (iv) the ordering of contiguous, large clones such as YAC and cosmid inserts for the

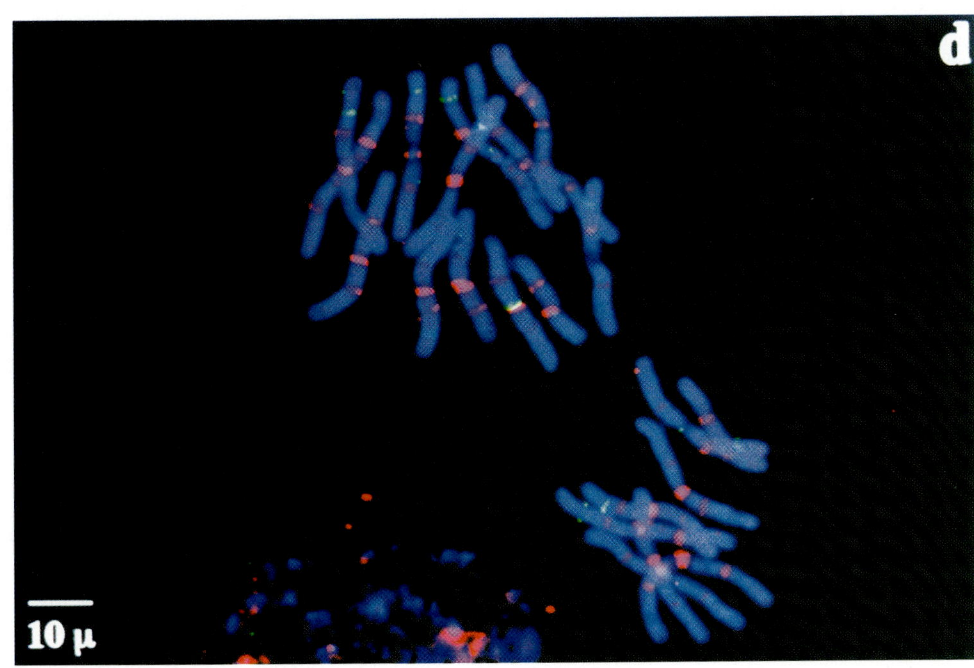

Fig. 5. In situ hybridization of rDNA, telomeric repeats and fluorochrome banding to mitotic metaphase chromosomes of root-tips of seedlings of slash pine: (a) in situ hybridization using a concatenated oligonucleotide telomeric repeat probe, detected red, showing the major and minor sites of telomeric repeats; (b) in situ hybridization showing the sites of 18S-25S rDNA detected green; (c) DAPI staining showing the chromosomes and bright, AT-enriched bands at most centromeres; (d) overlay of images a, b and c.

isolation of genes located on RFLP and RAPD linkage maps; and perhaps (v) enable analysis of chromosomal recombination in lines derived from interspecific hybrids.

Dense maps based on DNA markers can be constructed by segregation analysis. However, in high-resolution genetic linkage mapping, the smaller the map distance between loci, the larger the segregating population needed to ensure, at an acceptable level of probability, the proper ordering of loci. Large numbers of segregating individuals from controlled crosses may not be available, because of lack of polymorphisms or the time and space required to generate families, whereas to physically map loci requires seed from only one individual. With in situ hybridization, several probes can be detected independently and ordered along chromosomes in a single experiment, potentially making genome mapping a quicker and more efficient process than segregation analysis (Brown et al., 1993). Combining the two techniques of genetic mapping, Mendelian recombination and cytogenetic localization, holds the most promise for breeders and pathologists to address the immediate problem of fusiform rust disease, and more broadly the genetics of disease development. Increasing fundamental knowledge on the genetics of the host-pathogen interaction and linked markers in the pine will permit breeders to monitor populations for resistance genes improving longer-term breeding strategies.

I thank R. C. Schmidtling, T. L. Kubisiak, and A. Kamm for their reviews of earlier versions of this manuscript. I thank J. S. Heslop-Harrison for assistance in preparation of the final manuscript.

REFERENCES

Ahuja, M. R., Devey, M. E., Groover, A. T., Jermstad, K. D. and Neale, D. B. (1994). Mapped DNA probes from loblolly pine can be used for restriction fragment length polymorphism mapping in other conifers. *Theor. Appl. Genet.* **88**, 279-282.

Blackburn, E. H. (1991). Structure and function of telomeres. *Nature* **350**, 569-573.

Boyer, J. N. and South, D. B. (1984). Forest nursery practices in the South. *S. J. Appl. Forestry* **8**, 67-75.

Brown, G. R., Amarasinghe, V., Kiss, G. and Carlson, J. E. (1993). Preliminary karyotype and chromosomal localization of ribosomal DNA sites in white spruce using fluorescence in situ hybridization. *Genome* **36**, 310-316.

Byram, T. D., Greene, T. A., Lowe, W. J., McKinley, C. P., Robinson, J. F. and van Buijtenen, J. P. (1992). *Fortieth Progress Report of the Cooperative Forest Tree Improvement Program.* Forest Genetics Laboratory, Texas Forest Service Circular 292. Texas Forest Service, Texas A&M University, College Station, TX.

Cheung, W. Y., Money, T. A., Abbo, S., Devos, K. M., Gale, M. D. and Moore, G. (1994). A family of related sequences associated with (TTTAGGG)n repeats are located in the interstitial regions of wheat chromosomes. *Mol. Gen. Genet.* **245**, 349-354.

Conkle, M. T. (1981). Isozyme variation and linkage in six conifer species. In *Proceedings of the Symposium on Isozymes of North American Forest Trees and Forest Insects, July 1979, Berkeley, CA* (technical coordinator, M. T. Conkle), pp. 11-17. USDA Forest Service General Technical Report PSW-48.

Cox, A. V., Bennett, S. T., Parokonny, A. S., Kenton, A., Callissmassia, M. A. and Bennett, M. D. (1993). Comparison of plant telomere locations using a PCR-generated synthetic probe. *Ann. Bot.* **72**, 239-247.

Dhillon, S. S. (1980). Nuclear volume, chromosome size and DNA content relationships in three species of *Pinus. Cytologia* **45**, 555-560.

Doudrick, R. L. and Nelson, C. D. (1993). Complementary genetic interaction in fusiform rust disease on slash pine (Abstract). *Phytopathology* **83**, 1415.

Doudrick, R. L., Heslop-Harrison, J. S., Nelson, C. D., Schmidt, T., Nance, W. L. and Schwarzacher, T. (1995). The karyotype of *Pinus elliottii* var. *elliottii* using patterns of fluorescent in situ hybridization and fluorochrome banding. *J. Hered.* **86**, 289-296.

Gerlach, W. L. and Bedbrook, J. R. (1979). Cloning and characterization of ribosomal RNA genes from wheat and barley. *Nucl. Acids Res.* **7**, 1869-1885.

Greilhuber, J. (1988). Critical reassessment of DNA content variation in plants. In *Proceedings of the Third Chromosome Conference, Royal Botanic Gardens, Kew, England* (ed. P. E. Brandham), pp. 39-50. Royal Botanic Gardens, Kew, UK.

Griggs, M. M. and Walkinshaw, C. H. (1982). Diallel analysis of genetic resistance to *Cronartium quercuum* f. sp. *fusiforme* in slash pine. *Phytopathology* **72**, 816-818.

Guries, R. P., Friedman, S. T., Ledig, F. T. (1978). A megagametophyte analysis of genetic linkage in pitch pine (*Pinus rigida* Mill.). *Heredity* **40**, 309-314.

Haldane, J. B. S. (1919). The combination of linkage values and the calculation of distance between the loci and linked factors. *J. Genet.* **8**, 299-309.

Heslop-Harrison, J. S., Schwarzacher, T., Anamthawat-Jónsson, K., Leitch, A. R., Shi, M. and Leitch, I. J. (1991). In situ hybridization with automated chromosome denaturation. *Technique* **3**, 109-116.

Hizume, M., Ohgiku, A. and Tanaka, A. (1989). Chromosome banding in the genus *Pinus* II. Interspecific variation of fluorescent banding patterns in *P. densiflora* and *P. thunbergii. Bot. Mag. Tokyo* **102**, 25-36.

Hizume, M., Arai, M. and Tanaka, A. (1990). Chromosome banding in the genus *Pinus* III. Fluorescent banding patterns of *P. luchuensis* and its relationships among the Japanese diploxylon pines. *Bot. Mag. Tokyo* **103**, 103-111.

Ijdo, J. W., Wells, R. A., Baldini, A. and Reeders, S. T. (1991). Improved telomere detection using a telomere repeat probe (TTAGGG)$_n$ generated by PCR. *Nucl. Acids Res.* **19**, 4780.

Kinloch, B. B. and Walkinshaw, C. H. (1991). Resistance to fusiform rust: how is it inherited? In *Proceedings IUFRO Working Party Conference on Rusts of Hard Pines, 18-22 September 1989, Banff, Alberta* (ed. Y. Hiratsuka, J. K. Samoil, P. V. Blenis, P. E. Crane and B. L. Laishley), pp. 219-228. Forestry Canada Information Report NOR-317.

Kubisiak, T. L., Nelson, C. D., Nance, W. L. and Stine, M. (1995). RAPD linkage mapping in a longleaf pine × slash pine F₁ family. *Theor. Appl. Genet.* **90**, 1119-1127.

Lander, E. S., Green, P., Abrahamson J., Barlow, A., Daly, M. J., Lincoln, S. E. and Newburg, L. (1987). MAPMAKER: an interactive computer package for constructing primary genetic linkage maps of experimental and natural populations. *Genomics* **1**, 174-181.

Lincoln, S. E., Daly, M. J. and Lander, E. S. (1992). *Constructing Genetic Maps with MAPMAKER/EXP 3.0.* Whitehead Institute, Technical Report, 3rd edn. Whitehead Institute, Cambridge, MA.

Lohrey, R. E. and Kossuth, S. V. (1990). Slash pine. In *Silvics of North America. Volume 1, Conifers* (ed. R. M. Burns and B. H. Honkola), pp. 338-347. Agriculture Handbook 654, USDA Forest Service, Washington, DC.

MacPherson, P. and Filion, W. G. (1981). Karyotype analysis and the distribution of constitutive heterochromatin in five species of *Pinus. J. Hered.* **72**, 193-198.

Mehra, P. N. and Khoshoo, T. N. (1956). Cytology of conifers I. *J. Genet.* **54**, 165-180.

Mergen, F. (1958). Natural ploidy in slash pine. *Forest Sci.* **4**, 283-295.

Michelmore, R. W., Paran, I. and Kesseli, R. V. (1991). Identification of markers linked to disease-resistance genes by bulked segregant analysis: a rapid method to detect markers in specific genomic regions using segregating populations. *Proc. Nat. Acad. Sci. USA* **88**, 9828-9832.

Michelozzi, M., Squillace, A. E. and White, T. L. (1990). Monoterpene composition and fusiform rust resistance in slash pine. *Forest Sci.* **36**, 470-475.

Moran, G. F., Bell, J. C. and Hilliker, A. J. (1983). Greater meiotic recombination in male vs. female gametes in *Pinus radiata. J. Hered.* **74**, 62.

Mullin, L. J., Barnes, R. D. and Prevôst, M. J. (1978). A review of the southern pines in Rhodesia. The Rhodesia (Zimbabwe) Bulletin of Forestry Research No. 7.

Nance, W. L. and Schumate, C. B. (1992). *Automation of Random Amplified Polymorphic DNA* (RAPD) *Reactions.* Microlab® AT and AT*Plus* Application, Document L50035, Hamilton Co., Reno, Nevada. pp. 4.

Nance, W. L., Tuskan, G. A., Nelson, C. D. and Doudrick, R. L. (1992). Potential applications of molecular markers for genetic analysis of host-pathogen systems in forest trees. *Can. J. Forest Res.* **22**, 1036-1043.

Nelson, C. D., Doudrick, R. L., Nance, W. L., Hamaker, J. M. and Capo, B. (1993a). Specificity of host:pathogen genetic interaction for fusiform rust disease on slash pine. In *Proceedings of 22nd Southern Forest Tree Improvement Conference, June 1993, Atlanta Georgia* (ed. C. W. Lantz and J. Moorhead), pp. 403-410. Sponsored Publication No. 44, Southern Forest Tree Improvement Committee, National Technical Information Service, Springfield, Virginia.

Nelson, C. D., Nance, W. L. and Doudrick, R. L. (1993b). A partial genetic linkage map of slash pine (*Pinus elliottii* Engelm. var. *elliottii*) based on random amplified polymorphic DNAs. *Theor. Appl. Genet.* **87**, 145-151.

Nelson, C. D., Kubisiak, T. L., Stine, M. and Nance, W. L. (1994). A genetic linkage map of longleaf pine (*Pinus palustris* Mill.) based on random amplified polymorphic DNAs. *J. Hered.* **85**, 433-439.

Ogoshi, Y. and Nakata, G. (1955). Chromosome numbers of some species of Coniferae (Abstract). *Jap. J. Genet.* **30**, 186.

Ohri, D. and Khoshoo, T. N. (1986). Genome size in gymnosperms. *Plant Systemat. Evol.* **153**, 119-132.

Pan, Z. G. (1989). *Provenance Test of Slash and Loblolly Pine in China - 8th-Year Results.* Research Institute of Forestry, Chinese Academy of Forestry, Beijing, China.

Pederick, L. A. (1967). The structure and identification of the chromosomes of *Pinus radiata* D. Don. *Silvae Genetica* **16**, 69-77.

Picchi, C. G. and Barrett, W. H. C. (1967). Efecto de heladas intensas sobre las especies de *Pinus* cultivadas in Castelar. *IDIA: Suppl. Forestal* **4**, 1-11.

Rhoades, M. M. (1941). Different rates of crossing over in male and female gametes of maize. *J. Am. Soc. Agron.* **33**, 603-615.

Sax, K. and Sax, H. J. (1933). Chromosome number and morphology in the conifers. *J. Arnold Arboretum* **15**, 356-375.

Schmidt, T., Schwarzacher, T. and Heslop-Harrison, J. S. (1994). Physical mapping of rRNA genes by fluorescent in situ hybridization and structural analysis of 5s rRNA genes and intergenic spacer sequences in sugar beet (*Beta vulgaris*). *Theor. Appl. Genet.* **88**, 629-636.

Schwarzacher, T. and Heslop-Harrison, J. S. (1991). In situ hybridization to telomeres using synthetic oligomers. *Genome* **34**, 317-323.

Snow, G. A., Dinus, R. J. and Kais, A. G. (1975). Variation in pathogenicity of diverse sources of *Cronartium fusiforme* on selected slash pine families. *Phytopathology* **65**, 170-175.

Tulsieram, L. K., Glaubitz, J. C., Kiss, G. and Carlson, J. E. (1992). Single-tree genetic linkage mapping in conifers using haploid DNA from megagametophytes. *Bio/Technology* **10**, 686-690.

USDA Forest Service (1994). *Forest Insect and Disease Conditions in the United States, 1993.* pp. 73. United States Department of Agriculture, Forest Service, Forest Pest Management, Washington, D.C.

Wakamiya, I., Newton, R. J., Johnston, J. S. and Price, H. J. (1993). Genome size and environmental factors in the genus *Pinus. Am. J. Bot.* **80**, 1235-1241.

Welsh, J. and McClelland, M. (1990). Fingerprinting genomes using PCR with arbitrary primers. *Nucl. Acids Res.* **18**, 7213-7218.

Williams, J. G. K., Kubelik, A. R., Livak, K. J., Rafalski, J. A. and Tingey, S. V. (1990). DNA polymorphisms amplified by arbitrary primers are useful as genetic markers. *Nucl. Acids Res.* **18**, 6531-6535.

Zakian, V. A. (1989). Structure and fusion of telomeres. *Annu. Rev. Genet.* **23**, 579-604.

Zobel, B., Blair, R. and Zoerb, M. (1971). Using research data - disease resistance. *J. Forestry* **69**, 486-489.

Unifying plant molecular data and plants

Niels Jacobsen and Marian Ørgaard

Department of Botany, Dendrology and Forest Genetics, Royal Veterinary and Agricultural University, Rolighedsvej 23, DK-1958 Frederiksberg C (Copenhagen), Denmark

SUMMARY

Located at a botanical department at an Agricultural University, our taxonomical and genetic research is mainly directed towards cultivated plants and their wild relatives. The investigations are usually under a common heading 'experimental taxonomy', and include basic systematics, cytogenetics, biodiversity, population dynamics, conservation and evolutionary questions correlating the wild species and the cultivated forms. Our point of initiation is the plants and questions/problems raised regarding these plants. Our way of approaching the problems is usually by applying different sets of data and testing them. Experimental taxonomy covers classical cytogenetics (chromosome counting and karyotyping) as well as molecular cytogenetic methods (RAPD, RFLP, in situ hybridization), and includes also chemical data on isoenzymes and anthocyanins. We have had good collaborations with other laboratories and found their expertise on the plants in question very helpful. The aim is always to unify various data on the same set of problems, in order to get a more complete understanding of the plants. At present the department is working on several, quite different plant genera, comprising herbs, aquatic plants, and trees. The methods vary, depending on the plants and the problems in question. Some of the current investigations concern the horticultural genera *Lilium* and *Crocus*, in which the main point of interest is the study of chromosome evolution using fluorescence in situ hybridization; preliminary investigations into the composition of anthocyanins in *Crocus* look very promising. In the tropical starch tuber crop *Pachyrhizus* (Fabaceae), molecular analyses of relationships between existing cultivars, landraces and wild material have been carried out. A genus which we, in cooperation with a number of other laboratories, have been working with for many years is *Hordeum* (Poaceae) with one cultivated species (barley) and 31 wild species. Here the main areas of investigation have been field studies and collecting, followed by a taxonomical treatment, hybridization experiments, cytogenetic analysis and isoenzyme studies. Within the field of forestry, we have used population genetics as a tool in the management of natural and domesticated populations and for conservation of genetic diversity. We have also ventured into the identification and use of DNA markers that are suited for genome mapping in *Picea abies* (Norway Spruce).

Key words: Evolution, Triploid pathway, *Hordeum*, *Crocus*, *Lilium*, *Pachyrhizus*, *Abies*, *Picea*

INTRODUCTION

The investigation of plant molecular data has accelerated at a tremendous speed during the last decade. The number of methods used and the amount of data has reached a level where it has become yet more difficult to grasp the whole picture, even within only limited areas or even single plant genera or species. The amount of work involved in this kind of research tends to take up so much time that it is difficult to have time left to keep up with what is happening in adjacent fields of research, related plants or related problems of research. A result of this escalation is that the research often tends to forget the initial question, as one is buried in new methods and problems. The promising new perspectives tend to keep you moving into new exciting fields, with the result that you may forget to pick up and store the data that you, or someone else, actually needs, in order to retrieve it at a later date or make them reproducible.

Being within the field of experimental taxonomy, the problem of how to document results and how to make them available to others has come up a number of times. An even greater problem is how to make results reproducible, and how to store them for the future. A classical plant taxonomist solves this problem in a very convenient way: a herbarium specimen is preserved and placed in a museum designed for that purpose. Experimental taxonomists have also learned to store vouchers of the plant material used, and to document the origin etc., thereby making it possible for others to continue the research on the same or similar material. In the following, some examples from our department will be mentioned. Hopefully they will illustrate our basic approach to studying plant speciation, diversity, and variability.

HORDEUM

Our studies of the wild species of *Hordeum* began some twenty years ago, not so far off from the beginning of the 'molecular era'. Diter von Wettstein, at the Carlsberg Laboratory, working on the genetics of barley, emphasized the importance for the future genetics, molecular biology, and breeding work of barley, of establishing an all-round, up to date knowledge of the whole genus *Hordeum*, including taxonomy (32 species and 45 taxa altogether), distribution, variation, crossing ability, cytology, cytogenetics and cDNA (for references see e.g. von Bothmer et al., 1995; Jacobsen and von Bothmer, 1992; Linde-Laursen et al., 1992). Isoenzyme variation has also been used to unravel relationships between the different species, although it cannot be considered a phylogenetic reconstruction of the evolution (Jørgensen, 1986). A number of intergeneric crosses have been made (e.g. von Bothmer and Jacobsen, 1989; Salomon et al., 1991) and most recently Ørgaard and Heslop-Harrison (1993, 1994) have studied intergeneric hybrids between various species of *Hordeum* and *Leymus* by GISH and found it possible to discriminate the genomes of the two genera. Doebley et al. (1992) have studied the cDNA variation and the phylogeny of the genus *Hordeum*.

The basis for this work is the large number of accessions of wild species collected throughout the area of distribution, and now placed at the Nordic Gene Bank. An account of the localities and other relevant data have been made. It is not possible for the gene bank to maintain all the accessions that we have made, and there is at present no need for this, as we have found that most of them presently maintain themselves very well in nature, where they are important and dominating elements in the natural vegetation.

Transferring genetic material from the wild species of *Hordeum* to the cultivated barley will require further manipulation, as all hybrids with *H. vulgare* are sterile; but there are indications that it will be possible in the future. Using addition lines of e.g. *H. chilense* in *Triticum aestivum* (see e.g. P. Hernandez and A. Martin, unpublished) is another way of transferring genetic material.

The extensive research done on the grass tribe Triticeae, which comprises about 25 genera related to the three cereals wheat (*Triticum*), barley (*Hordeum*), and rye (*Secale*), show that they all (?) constitute a potential genepool for these major crops. Not a primary genepool but rather a secondary or tertiary genepool (see e.g. von Bothmer et al., 1992).

Worth considering is that one of our aims in the studies of *Hordeum* was to construct a phylogenetic tree, thereby assuming that a tree-like evolution has taken place. What if this is not the case? A recent review article dealing with the flora of Malesia (Johns, 1995), describes how New Guinea, which is one of the areas of the world with the quickest uplift of land, has one of the most diverse floras found today. In these unstable regions the greatest diversification of species is concentrated. About 10% of all flowering plants of today are found on New Guinea, and for the part of the flora that has been investigated so far, it has been found that 50% of the Dicotyledons and 25% of the Monocotyledons are endemic for the island; a probable estimate for the whole flora is that 60-70% is endemic. It could be that an explosive, large scale evolution of plants has occurred during unstable geological and climatical conditions similar to those of New Guinea, when new environmental conditions are created at great speed. Under such unstable and new conditions, with constantly occurring mutations, plants may have the opportunity to radiate within much wider limits than usual. Under normal, stable ecological

conditions, only very few mutations survive more than a few generations unless they have great selective advantages. Could it be that 'big steps' in evolution take place under extreme, unstable conditions that facilitate the survival of mutations far beyond normal rates, and that an illustration of the evolution of the species we encounter today is more like phylogenetic fireworks than like regular, dichotomously branched, phylogenetic trees?

CROCUS

One of our recent lines of research concerns the well known European garden plants of the genus *Crocus* (c. 80 species distributed mostly in the Mediterranean region as far as eastern Turkey and Iran). In contrast to the genus *Hordeum*, which has fixed chromosome numbers (2n = 14, 28, 42), *Crocus* has a whole series of numbers: 2n = 6, 8 10, 13, 14, 16, 18, 20, 22, 23, 24, 25, 26, 27, 28, 30, 32, 34, 35, 36, and 64. Many hybrids are sterile, but a few are more or less fertile. Their taxonomy was studied by Mathew (1982), and the cytology mostly by Brighton (see e.g. Brighton et al., 1980). The aneuploid series found both within the genus as well as within the different *Crocus* species suggests a much more versatile system of chromosome behaviour than e.g. that found in *Hordeum* and other Triticeae. The suggestion of a 'triploid pathway' (Ørgaard et al., 1995b) for creating aneuploid series of chromosome numbers in plants may find support in a closer study of the genus.

Of special interest is the series Biflori, comprising among other species the complex *C. biflorus* (15 subspecies, and 2n = 8, 10, 12, 16, 18, 20, 22) and *C. chrysanthus* (2n = 8, 10, 12, 13, 14, 16, 18, 19, 20). The hybrids between the two are the early, relatively small-flowered cultivars, commonly used as garden plants.

A study of the triploid hybrid *Crocus* 'Golden Yellow' demonstrated that it is possible, using genomic in situ hybridization, to differentiate the parental genomes (Ørgaard et al., 1995a,b).

A preliminary study of flower pigments (anthocyanins) of *Crocus* has revealed that different species and subspecies have distinguishable anthocyanins, characteristics which may be very useful as markers for discriminating species and the parentage of hybrids. This venture into *Crocus* anthocyanins was triggered by a successful study of the anthocyanin components in flowers of the genus *Alstroemeria* (Alstroemeriaceae) by Nørbæk et al. (1996).

LILIUM

The genus *Lilium* has been used in cytological studies for many years because of its rather large chromosomes, normally 2n = 24 (for review see Noda, 1991). The genus contains about 80 species distributed in temperate (to subtropical) areas in the northern hemisphere (e.g. Synge, 1980). Compared to *Hordeum* and *Crocus*, *Lilium* represents a genus with a relatively stable chromosome compliment and the fertility of the interspecific hybrids is often high in F_1, F_2 and further generations. Results hitherto obtained by the use of genomic in situ hybridization (unpublished), indicate that it may be possible to trace the heritage of genomes, chromosomes, and chromosome segments in generations beyond the F_1 generation. Our main target plants are the old, *L.* × Testaceum (probably *L. chalcedonicum* × *L. candidum*, both from Europe), *L.* × Hollandicum (*L. bulbiferum* (Europe) × *L. maculatum* (Japan), the progenitor to the so called 'Asiatic hybrids' – the brightly coloured, commonly sold cut-flower lilies). In hybrid combinations: 'Asiatic hybrids' × *L. pumilum* (NE Asia), a rather high percentage of unreduced pollen gametes is found (see e.g. van Tuyl, 1990; van Tuyl and Kwakenbos, 1989). These unreduced 2n gametes are potential donors in the production of triploid plants, and thereby the study of the 'triploid pathway', possibly creating an aneuploid chromosome series (Ørgaard et al., 1995b). Besides using the produced *Lilium* hybrids for cytogenetic research purposes, the hybrid material will also be used in future breeding programmes. A number of the combinations were made in order to obtain genetic dwarf cultivar types for use as pot plants (i.e. a height of 30-40 cm), thus avoiding the use of growth retardents.

PACHYRHIZUS

Pachyrhizus or Yambean, from South to Central America, is a genus of about five species (Sørensen, 1990), widely cultivated (also in Asia and Africa) as a tuberous legume, which can be grown both on a small scale for local consumption and marketing, as well as on a larger scale as a cash crop. Reports from field trials claim more than 100 t/ha in Tonga (e.g. M. Grum et al., unpublished). The tuber is crisp and rich in starch, and can be transported as well as stored. A large proportion of the *P. tuberosus* grown in Mexico is now exported to the US for use as a component in salad bars. The beans contain a toxic substance Rotenone and are therefore not edible; in Asia, flour made from the tuber may be put into cookies to keep insects away.

The project, which is funded by the EU, started about ten years ago with extensive collecting, hybridization experiments, taxonomic treatment (Sørensen, 1990; Sørensen et al., 1993), followed by large scale prebreeding experiments. Previously almost no planned breeding had been done in the genus, and what had been done only included the Central American *P. erosus*. *P. tuberosus* (tropical South America) and *P. ahipa* (mainly Bolivia) are the two other species which have been cultivated and they have now been crossed with *P. erosus* in as many accession combinations as possible. This hybrid material has been carried through further generations, and distributed to various breeding stations in the Pacific, Asia, Africa, Central and South America, in order to find the best suited cultivars for specific regions (e.g. M. Grum et al., unpublished). At the University of St Andrews, Scotland, the *Pachyrhizus* material is being used in studies of cDNA variation, in order to elucidate relationships within the genus.

ABIES AND PICEA

The main goal at the Arboretum is to gather as many accessions of various species in order to have them under trial for use under Danish conditions. Within the area of forest genetics and forest tree improvement, the time scale is much greater.

This large collection of material has proven to be very suitable for genetical and molecular analysis.

Within *Abies* (not native to Denmark) isoenzyme analysis has been used to analyze the variation found in domesticated populations, and in investigation of the mating systems in a seed orchard of *A. procera* (Noble Fir). It has been found that the clones in the seed orchard are outcrossing, although the pollen pool was variable within the orchard, and that the effective population number was only 65% (Siegismund et al., 1996). When considering optimal sizes of gene pool nurseries in the breeding of landscape plants or tree seed production, it has been customary to use only few clones. This will result in inbreeding depressions when plants with only a limited genetic basis are flooded into these environments.

The Norway Spruce (*Picea abies*) is, like *Abies,* not native to Denmark, but it is widely cultivated and naturalized. A project on the identification of DNA markers well suited for genome mapping is almost completed and will enable us to search for linkage to gene loci coding for QTLs like health, growth, and quality. A partial genome map has been constructed comprising 12 linkage groups, based on 42 RAPD-markers (out of 105). The existence of a large female gametophyte (haploid) greatly facilitates the mapping of genes donated by the male and female gametophyte, respectively (E. Skov and H. Wellendorf, unpublished).

We thank our colleagues for sharing their unpublished work.

REFERENCES

Brighton, C. A., Scarlett, C. J. and Mathew, B. (1980). Cytological studies and origins of some *Crocus* cultivars. In *Petaloid Monocotyledons* (ed. C. D. Brickell, D. F. Cutley and M. Gregory), pp. 139-162. Linnean Society, Academic Press, London.

Doebley, J., von Bothmer, R. and Larson, S. (1992). Chloroplast DNA variation and the phylogeny of *Hordeum* (Poaceae). *Am. J. Bot.* **79**, 576-584.

Jacobsen, N. and von Bothmer, R. (1992). Supraspecific groups in the genus *Hordeum. Hereditas* **116**, 21-24.

Johns, R. J. (1995). Malesia – an introduction. *Curtis's Botan. Mag.* **12 (2)**, 52-62.

Jørgensen, R. B. (1986). Relationships in the barley genus (*Hordeum*). *Hereditas* **104**, 273-291.

Linde-Laursen, I., von Bothmer, R. and Jacobsen, N. (1992). Relationships in the genus *Hordeum. Hereditas* **116**, 111-116.

Mathew, B. (1982). *The* Crocus. B. T. Batsford Ltd, London.

Noda, S. (1991). Chromosomal variation and evolution in the genus *Lilium*. In *Chromosome Engineering in Plants: Genetics, Breeding, Evolution*, part B, ch. 27 (ed. P. Tsuchiya and P. K. Gupta), pp. 507-524.

Nørbæk, R., Christensen, L. P., Bojesen, G. and Brandt, K. (1996). Anthocyanins in Chilean species of *Alstroemeria. Phytochemistry* (in press).

Ørgaard, M. and Heslop-Harrison, J. S. (1993). Relationships between species of *Leymus, Psathyrostachys* and *Hordeum* (Poaceae, Triticeae) inferred from Southern hybridization of genomic DNA and cloned DNA probes. *Plant Syst. Evol.* **189**, 217-231.

Ørgaard, M. and Heslop-Harrison, J. S. (1994). Investigations of genome relationships in *Leymus, Psathyrostachys* and *Hordeum* by genomic DNA:DNA *in situ* hybridization. *Ann. Bot.* **73**, 195-203.

Ørgaard, M., Jacobsen, N. and Heslop-Harrison, J. S. (1995a). The hybrid origin of two *Crocus* (Iridaceae) analyzed by molecular cytogenetics including genomic Southern and *in situ* hybridization. *Ann. Bot.* **76**, 253-262.

Ørgaard, M., Jacobsen, N. and Heslop-Harrison, J. S. (1995b). Molecular cytogenetics in the genus *Crocus* L. In *Kew Chromosome Conference IV* (ed. P. E. Brandham and M. D. Bennett), pp. 291-299 + xxxv. Royal Botanic Gardens, Kew, London.

Salomon, B., von Bothmer, R. and Jacobsen, N. (1991). Intergeneric crosses between *Hordeum* and North American *Elymus* (Poaceae, Triticeae). *Hereditas* **114**, 35-39.

Siegismund, H. R., Kjær, E. D. and Nielsen, U. B. (1996). Mating system estimates for an isolated Noble Fir (*Abies procera*) clonal seed orchard in Denmark. *Can. J. For. Res.* (submitted).

Sørensen, M. (1990). Observations on distribution, ecology and cultivation of the tuber-bearing legume genus *Pachyrhizus* Rich. ex DC. *Wageningen Agric. Papers* **90**, **3**, 1-38.

Sørensen, M., Grum, M., Paull, R. E., Vaillant, A., Venthou-Dumaine and Zinsou, C. (1993). Yam bean (*Pachyrhizus* species). In *Underutilized Crops. 1. Pulses and Vegetables*, ch. 2 (ed. J. T. Williams), pp. 59-102.

Synge, P. M. (1980). *Lilies*. B. T. Batsford Ltd, London.

van Tuyl, J. M. and Kwakenbos, T. A. M. (1989). Research on polyploidy in interspecific hybridization of lily. *Lily Yearbook, North American Lily Society* **42**, 62-65.

van Tuyl, J. M. (1990). Survey research on mitotic and meiotic polyploidization of CPRO-DLO. *Lily Yearbook, North American Lily Society* **43**, 10-13.

von Bothmer, R. and Jacobsen, N. (1989). Intergeneric hybridization between *Hordeum* and *Hordelymus* (Poaceae). *Nord. J. Bot.* **9**, 113-117.

von Bothmer, R., Seberg, O. and Jacobsen, N. (1992). Genetic resources in the Triticeae. *Hereditas* **116**, 141-150.

von Bothmer, R., Jacobsen, N., Baden, B., Jørgensen, R. B. and Linde-Laursen, I. (1995). An ecogeographical study of the genus *Hordeum*. In *Systematic and Ecogeographical Studies on Crop Genepools 7*, 2nd edn, pp. 1-129. International Plant Genetic Resources Institute, Rome.

Printed in Great Britain © The Society for Experimental Biology 1996
SEB0031

Molecular characterization of heterochromatic regions around the *Tm-2* locus in chromosome 9 of tomato

Fusao Motoyoshi, Taku Ohmori and Minoru Murata

Research Institute for Bioresources, Okayama University, Kurashiki 710, Japan

SUMMARY

The gene *Tm-2* (tomato mosaic virus (ToMV) resistant), which is tightly linked to a morphological marker gene *nv* (netted virescent), resides in a heterochromatic region near the centromere of chromosome 9 in tomato. *Tm-2* and *Tm-2a* are known to be allelic, and exhibit similar phenotypes to each other, but can be differentiated by their response to different ToMV strains. An inoculation experiment demonstrated that *Tm-2* helped a mutant strain of ToMV to infect a heterozygous tomato (*Tm-2/Tm-2a*). Aiming at investigating the structures of DNA around these active genes and their influence on gene activities, we attempted to identify and characterize random amplified polymorphic DNA (RAPD) markers linked to these genes using nearly isogenic lines (NILs) of tomato. Genetic analysis using 13 RAPD markers linked to the *Tm-2* locus and cytological analysis by fluorescence in situ hybridization (FISH) demonstrated that the lines resistant to ToMV had a large block derived from a chromosome of *Lycopersicon peruvianum*. Among these markers, we estimated that two, $OPE16_{900}$ and $OPN31_{1000}$, are nearest to the *Tm-2* locus. Out of the 13 markers six, distributed within about 0.7 centi-Morgan (cM), were cloned and sequenced to be converted to sequence characterized amplified region (SCAR) markers. Of these, four were successfully converted to SCAR markers. The six clones were also used as probes for Southern hybridization of genomic DNA from NILs to characterize structures around the *Tm-2* locus. One clone was estimated to be derived from a sequence that was present in one copy. The other five clones appeared to be derived from different kinds of moderately or highly repetitive sequences.

Key words: Heterochromatin, RAPD, SCAR, *Tm-2*, Tomato mosaic virus (ToMV), FISH

INTRODUCTION

Chromosomes of some higher plants have heterochromatin surrounding their centromeres. The structure and function of this heterochromatin is only poorly understood. In tomato (*Lycopersicon esculentum*) chromosomes, typical large heterochromatic regions exist around the centromeres. These heterochromatic regions have been presumed to lack gene activity (Snoad, 1963) with some exceptions. As a rare case, Khush et al. (1964) demonstrated that *nv* and *Tm-2* reside near the centromere in the heterochromatic region of chromosome 9. *Tm-2a* must have its locus in the heterochromatic region, since it is genetically allelic to *Tm-2* (Pecaut, 1965). It is probable that the loci of these active genes themselves are not heterochromatic, though they are surrounded by heterochromatin (Khush et al., 1964).

We are interested in what molecular structures the heterochromatin around active genes consists of, and whether the heterochromatic region influences the expression of active genes that reside there. To provide an answer to these problems, we are attempting to characterize a heterochromatic region near the centromere of chromosome 9 in tomato. In a previous study, we mapped the genes *nv*, *Tm-2* and *Tm-2a* with genetically linked random amplified polymorphic DNA (RAPD) markers (Ohmori et al., 1995). We describe some characteristics of the genes, *nv*, *Tm-2* and *Tm-2a*, and some molecular properties of clones isolated from RAPD fragments linked to these genes.

GENE ACTIVITIES IN THE HETEROCHROMATIC REGION NEAR THE CENTROMERE IN CHROMOSOME 9 IN TOMATO

Cotyledons and true leaves of plants homozygous for the recessive gene *nv* are pale green in color in their interveinal areas (Clayberg et al., 1960). Plants homozygous for *nv* are stunted specifically at their young stages, but grow until ripening. Because of very tight linkage with *Tm-2* (Soost, 1963), which had been introgressed from *L. peruvianim* into tomato, the locus of *nv* may have been derived also from this wild relative species of tomato. Neither biochemical nor molecular biological analyses of the expression of *nv* and its wild-type allele have been made, as far as we know.

Tm-2 and *Tm-2a* express a similar resistance to wild-type ToMV strains, but can be differentiated with respect to their responses to some mutant ToMV strains. For example, tomato seedlings whose genotype is either *Tm-2/Tm-2* or *Tm-2/+* are severely infected with a ToMV mutant strain Ltb1, which was derived from a wild-type strain L (Ohno et al., 1984), but those

Table 1. Expression of *Tm-2* and *Tm-2a* genes in response to ToMV strains L and Ltb1 (9 days after inoculation)

| Leaf | Combination in F1 (genotype) | No. of local lesions (4 leaves) | | |
		ToMV strain	Sample/ standard	Ratio
Cotyledon (inoculated)	GCR267×GCR26 (*Tm-2a/+*)	L	31/315	0.1
		Ltb1	2/290	0.001
	GCR236×GCR26 (*Tm-2/+*)	L	12/248	0.05
		Ltb1	814/281	2.9
	GCR236×GCR267 (*Tm-2/Tm-2a*)	L	20/404	0.05
		Ltb1	1098/265	4.1
Third true leaf (systemic)	GCR267×GCR26 (*Tm-2a/+*)	L	0/304	0
		Ltb1	0/402	0
	GCR236×GCR26 (*Tm-2/+*)	L	6/294	0.02
		Ltb1	770/251	3.1
	GCR236×GCR267 (*Tm-2/Tm-2a*)	L	0/345	0
		Ltb1	894/256	3.5

of either *Tm-2a/Tm-2a* or *Tm-2a/+* are still resistant to this ToMV strain (Meshi et al., 1989).

Moreover we found that seedlings with a genotype *Tm-2/Tm-2a* are susceptible to Ltb1, indicating that in this genotype the *Tm-2* gene confers susceptibility, and behaves as though it is dominant to the resistance by *Tm-2a*. In one experiment (Table 1), F$_1$ plants obtained by crossing of GCR236 (*Tm-2/Tm-2*, *nv/nv*) with GCR26(+/+, +/+) and those obtained by crossing of GCR236(*Tm-2a/Tm-2a*, +/+) with GCR26(+/+, +/+) were compared to other F$_1$ plants obtained by crossing of GCR236(*Tm-2/Tm-2*, *nv/nv*) with GCR 267(*Tm-2a/Tm-2a*, +/+), with respect to their response to ToMV strains, L and Ltb1. Cotyledons of 8-day-old plants were inoculated with ToMV L or Ltb1 at a concentration of 10 μg of purified virus/ml of 0.01 M Na-phosphate buffer, pH 7.0, and grown in a temperature-controlled greenhouse at about 25°C. At 9 days after inoculation, the inoculated cotyledons and the third true leaves were harvested. To assay the infectivity of virus accumulated in the inoculated cotyledons, and of systemically spread and accumulated virus in the third true leaves, the cotyledons and the true leaves were homogenized with 1 ml of 0.01 M Na-phosphate buffer, pH 7.0, per 1 g fresh weight, and diluted 5,000 times with the same buffer. Each sample was inoculated to four half-leaves of Xanthi nc tobacco, and a purified wild-type virus at the concentration of 0.02 μg/ml of the same buffer was inoculated to the opposite half of each leaf as standard. As a result, growth of Ltb1 was suppressed by *Tm-2a*, but in the presence of *Tm-2* the mutant virus can grow rapidly. Moreover, *Tm-2* appears to promote multiplication and systemic spreading of Ltb1 in the genotype *Tm-2/Tm-2a*.

The resistance expressed by the *Tm-2* gene involves an alteration in, or inhibition of the function of the virus-encoded 30 kDa protein that is necessary for spread of the virus from cell to cell (Meshi et al., 1989). The *Tm-2a* gene may have the same effect as *Tm-2* (Nishiguchi and Motoyoshi, 1987). The mechanism by which *Tm-2* helps Ltb1 virus infection in the presence of *Tm-2a* is not yet understood.

RAPD MARKERS LINKED WITH THE *nv* AND *Tm-2* LOCI

The molecular structure of heterochromatin near the cen-tromeres in tomato has been scarcely characterized so far. A tandemly repeated 162 bp satellite DNA that is mostly clustered at or near the telomeres and an interspersed repeat, which is referred to as TGRIII, were reported by Ganal et al. (1988) to be also present at or near the centromeres of some chromosomes.

In *Arabidopsis thaliana*, tandem repeat sequences of a 180 bp unit are known to be localized at or near centromeres (Murata et al., 1994). We examined by Southern hybridization of total DNA from tomato (GCR236 line) whether it has sequences similar to the 180 bp repeats. The tomato DNA (5 μg) was digested with *Bam*HI, *Ban*III, *Dra*I, *Eco*RI, *Eco*RV, *Hind*III, *Kpn*I, *Pst*I, *Sac*II, *Sal*I, *Xba*I or *Xho*I and electrophoresed in a 2% agarose gel. As a control, 2 μg of total DNA from *A. thaliana* (ecotype WS) was digested with *Hind*III and electrophoresed. The result of the Southern hybridization showed that no signal was observed in any lane of tomato DNA, in contrast to the ladder-like strong signals of the *A. thaliana* sample, suggesting that sequences homologous to the 180 bp repeat sequence of *A. thaliana* are not present in the tomato genome.

In tomato, therefore, there has been little information about the molecular structure of heterochromatin near the centromere. In chromosome 9, however, there are genes, *nv*, *Tm-2* and *Tm-2a*, in the heterochromatic area, which may be useful keys to collect DNA markers from the heterochromatin and, in turn, such DNA markers may become tools to investigate the structure of the heterochromatin. A high proportion of DNA markers, such as restriction fragment length polymorphism (RFLP) and random amplified polymorphic DNA (RAPD) markers, genetically linked to the *nv* as well as the *Tm-2* locus, may be useful markers for analysing heterochromatic DNA, where recombination is highly suppressed (Rick, 1971; Ganal et al., 1989).

We previously reported that 53 RAPDs were found among genomic DNA samples from three nearly isogenic lines (NILs) of tomato, GCR236, GCR267 and GCR26, by using 220 different 10 base oligonucleotide primers (OperonTM Technologies, Alamede, California) (Ohmori et al., 1995). GCR lines have a common genetic background from 'Craigella', a variety of tomato, but GCR236 carries homozygous *Tm-2* and *nv*, GCR267 carries homozygous *Tm-2a* but no *nv*, and GCR26 carries neither *Tm-2*, *Tm-2a*, nor *nv*. These lines were originally bred at the Glasshouse Crops Research Institute (Littlehampton, UK) (Smith and Ritchie 1983). Thirteen RAPD bands, which were arbitrarily selected, were shown to be linked to the *nv* locus within an interval of about 5.7 cM (Ohmori et al., 1995). Among these 13 RAPDs, nine markers were tightly linked to *nv* and to each other in the 108 BC$_1$ or 133 F$_2$ plants examined (Fig. 1).

It is known that centromeric heterochromatin suppresses recombination (Rick, 1971; Ganal et al., 1989). This means that the physical distance between genetically close markers is comparatively long. In a physical mapping study by Ganal et al. (1989), three RFLP markers within 1.2 cM on chromosome 9 were distributed over approximately 4 million bp. Our RAPD markers are therefore possibly distributed over 18 million bp. We found about 50 RAPD markers differentiating GCR26, GCR236 and GCR267. If all of them link to the *Tm-2* locus within a genetic distance of 5.4 cM, the distance between two neighboring markers is on average 360 kb.

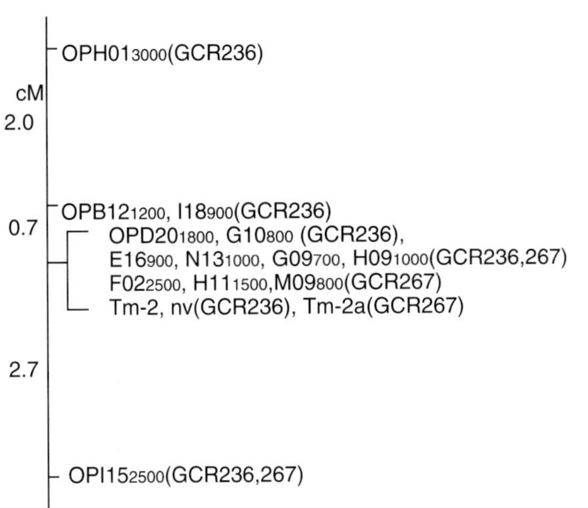

Fig. 1. A genetic map of RAPD markers together with genes *nv*, *Tm-2* and *Tm-2a*. The tomato lines in which RAPD markers are present are indicated in parenthesis.

It is desirable to determine what markers are nearer to and what markers are more distant from the *nv* as well as the *Tm-2* locus. As mentioned above, *Tm-2* and *Tm-2a* are allelic (Pecaut, 1965). It is thus probable that RAPD markers most close to the *Tm-2* locus are among those present in both GCR236 (*Tm-2/Tm-2*, *nv/nv*) and GCR267 (*Tm-2a/Tm-2a*, +/+). Under this assumption, markers closer to the *Tm-2* locus are OPE16$_{900}$, N13$_{1000}$, G09$_{700}$, H09$_{1000}$ and I15$_{2500}$. OPI15$_{2500}$, however, can be omitted, since this marker had been proved to be 2.7cM away from *nv* and *Tm-2* (Ohmori et al., 1995). In order to select the nearest marker(s) to the *Tm-2* locus, we compared a few tomato lines carrying *Tm-2* alleles with respect to the presence or absence of the RAPD markers in question (Table 2). The tomato line LA1000 is homozygous for *Tm-2*, *nv* and *ah* (Hoffman's anthocyaninless; about 2 cM away from *nv*). This line possessed all RAPD markers that were carried by GCR236. TMJ54-28 carries *Tm-2* type allele, which was introgressed from *L. peruvianum* P.I.126926 into tomato (Yamakawa and Nagata, 1975). This line was proved to lack G09$_{700}$ and H09$_{1000}$, but possess E16$_{900}$ and N13$_{1000}$. Perou2 carries another *Tm-2* type allele which was introduced from *L. peruvianum* P.I.126926 into tomato (Laterrot and Pecaut, 1969). OPE16$_{900}$ and N13$_{1000}$ were present but G09$_{700}$ and H09$_{1000}$ were absent in this line as observed in TMJ54-28. It is thus probable that E16$_{900}$ and N13$_{1000}$ are nearer to the *Tm-2* locus than the other markers examined.

CLONING OF SOME RAPD MARKERS LINKED TO THE *Tm-2* LOCUS

Six RAPD markers linked to the *Tm-2* locus were successfully cloned using *Sma*I-digested pBluescript II (KS+) vector (Stratagene). The cloned RAPD markers were OPB12$_{1200}$, I18$_{900}$ and G10$_{800}$, which were linked to the *Tm-2* gene (and the *nv* gene), but not to the *Tm-2a* gene, and OPE16$_{900}$, G09$_{700}$ and N13$_{1000}$, which were linked to both the *Tm-2* (and *nv*) and the *Tm-2a*.

In order to ascertain whether the clones were derived correctly from DNAs of the RAPD markers, we used Southern hybridization with the clones as probes to the RAPD markers prepared from tomato lines GCR26 (+/+, +/+), GCR236 (*Tm-2/Tm-2*, *nv/nv*), GCR267 (*Tm-2a/Tm-2a*) and *L. peruvianum* (P.I.128650). As a result, the clone from OPG09$_{700}$ generated strong signals in the two ToMV resistant lines and *L. peruvianum*. This clone, however, generated a weak signal in the susceptible line GCR26, suggesting that DNA from this line has a sequence similar to but not identical with those found in the resistance lines. The clone corresponding to OPB12$_{1200}$, which was present only in GCR236 (*Tm-2/Tm-2*, *nv/nv*) did not hybridize to any bands in the susceptible line GCR26, but hybridized to the RAPD band in GCR236 as well as a band of similar size in GCR267. All of the other clones gave signals in the resistant lines, but not in the susceptible line. The results suggest that all of the clones have been derived from the RAPD markers. Moreover, it is possible that the sequences of these clones were originally present in *L. peruvianum*, from which they were introgressed into tomato.

CONVERSION OF THE RAPD MARKERS TO SEQUENCE CHARACTERIZED AMPLIFIED REGION (SCAR) MARKERS

SCAR markers are superior to RAPD markers because they are more easily identified, usually as a single distinct band in an agarose gel (Paran and Michelmore, 1993; Adam-Blondon et al., 1994; Maisonneuve et al., 1994). Nucleotide sequences at both ends of five out of the six RAPD clones were determined and five sets of SCAR primers with 24 nucleotide sequences, which include 10 nucleotide sequences of the RAPD primers,

Table 2. Presence or absence of RAPD markers in different tomato lines with or without ToMV resistance genes

RAPD marker linked to *nv* and *Tm-2*	Tomato line						
	GCR26 (+)	Ponderosa (+)	GCR236 (*Tm-2*)	LA1000 (*Tm-2*)	TMJ54-28 (*Tm-2* type)	Perou2 (*Tm-2* type)	*L. peruvianum* P.I. 128650
H01$_{3000}$(GCR236)	–	–	+	+	+	+	–
B12$_{1200}$(GCR236)	–	–	+	+	+	–	–
I18$_{900}$(GCR236)	–	–	+	+	+	+	–
D20$_{1800}$(GCR236)	–	–	+	+	+	+	+
G10$_{800}$(GCR236)	–	–	+	+	+	+	+
E16$_{900}$(GCR236,267)	–	–	+	+	+	+	+
N13$_{1000}$(GCR236,267)	–	–	+	+	+	+	+
G09$_{700}$(GCR236,267)	–	–	+	+	–	–	+
H09$_{1000}$(GCR236,267)	–	–	+	+	–	–	+
I15$_{2500}$(GCR236,267)	–	–	+	+	–	+	+

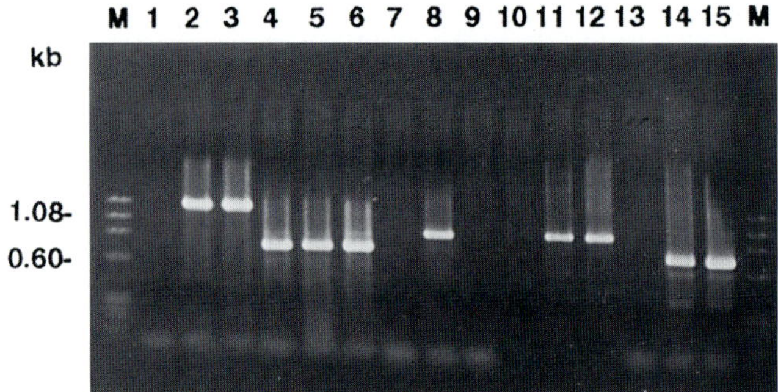

Fig. 2. Products amplified with five SCAR primers from genomic DNA samples in GCR lines. DNA samples were prepared from GCR26 (+/+, +/+) (lanes 1, 4, 7, 10 and 13), GCR236 (*Tm-2/Tm-2*, *nv/nv*) (lanes 2, 5, 8, 11 and 14), and GCR267 (*Tm-2a/Tm-2a*, +/+) (lanes 3, 6, 9, 12 and 15). The SCARs are designated $SCB12_{1200}$ (lanes 1-3), $SCG10_{800}$ (lanes 4-6), $SCI18_{900}$ (lanes 7-9), $SCE16_{900}$ (lanes 10-12) and $SCG09_{700}$ (lanes 13-15). Molecular size markers are a ϕX174/*Hae*III digest (lanes M).

were synthesized by means of an automated DNA synthesizer (391, Applied Biosystems). PCR was carried out by the method of Paran and Michelmore (1993) except that annealing was carried out at 65°C, and the amplified products were resolved by electrophoresis in a 1.4% agarose gel. A single band of the same size as the corresponding RAPD marker was detected with every set of primers. Among these five SCARs, $SCB12_{1200}$, $SCE16_{900}$, $SCI18_{900}$ and $SCG09_{700}$, corresponding to $OPB12_{1200}$, $OPE16_{900}$, $OPI18_{900}$ and $OPG09_{700}$, respectively, were present in GCR236 (*Tm-2/Tm-2*, *nv/nv*) as well as GCR267 (*Tm-2a/Tm-2a*, +/+), but not in GCR26 (+/+, +/+), whereas a same-sized band corresponding to $OPG10_{800}$ was present also in GCR26 (Fig. 2). Digestion of this PCR product with *Hae*III, *Hap*II or *Mbo*I did not show polymorphisms among the three tomato lines.

CHARACTERIZATION OF THE REGION AROUND THE *Tm-2* LOCUS

Table 3 shows that the RAPD markers, which were present in GCR267 (*Tm-2a/Tm-2a*) and confirmed to link tightly to *Tm-2a*, were all also kept in two commercial varieties Zuiei and Paster-Yoozu (Sakata Seed Co.) carrying this ToMV-resistance gene, while a ToMV-susceptible cultivar 'Ponderosa' did not have any of these eight RAPD markers. These RAPD markers exist also in *L. peruvianum* P.I.128650 (Table 2) which the

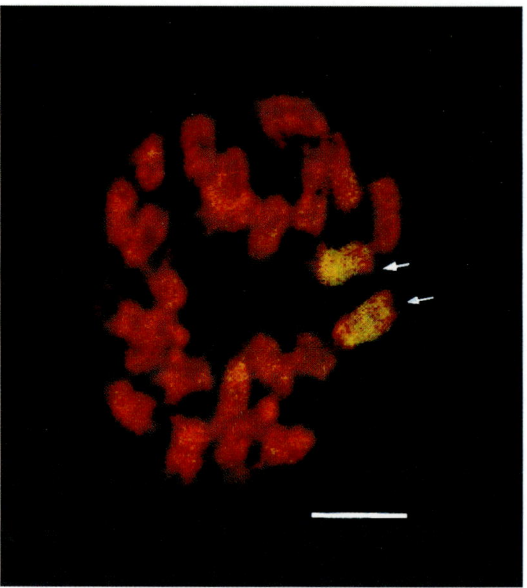

Fig. 3. Mitotic metaphase chromosomes of a tomato line GCR236 (*Tm-2/Tm-2*, *nv/nv*). The chromosomes were hybridized with a 200:1 mixture of DNA prepared from GCR26 (+/+, +/+) and biotinylated DNA from *Lycopersicon peruvianum* P.I.128650. FITC-avidin was applied to the preparation to detect hybridized probe. The chromosomes stained with propidium iodide were analyzed using a confocal lazer microscope (Bio-Rad MRC600). Arrows indicate chromosomes with FITC-signals that have possibly originated from *L. peruvianum*.

Tm-2a originated from (Alexander, 1963). The *Tm-2* gene was derived from another line of *L. peruvianum*, P.I.126926. We could not determine whether all the RAPD markers present in GCR236 (*Tm-2/Tm-2*) exist also in *L. peruvianum* P.I.126921, because we did not have the seeds of this line. Seven out of 10 RAPD markers are, however, shared between GCR236 and *L. peruvianum* P.I.128650, the donor of the *Tm-2a* gene. It is thus possible that *Tm-2* also exists in a region of considerable length that has been introgressed from *L. peruvianum* into GCR236. In a fluorescence in situ hybridization (FISH) study of tomato chromosomes using biotinylated genomic DNA from *L. peruvianum* as a probe, large blocks in a pair of chromosomes in GCR236 (*Tm-2/Tm-2*) as well as in GCR267 (*Tm-2a/Tm-2a*) showed FITC-signals, suggesting that these blocks were derived from *L. peruvianum* (Fig. 3).

In the next step, we carried out Southern hybridization with the six RAPD clones as probes to restriction endonuclease

Table 3. Presence or absence of RAPD markers in different tomato lines or varieties with or without *Tm-2a* gene

| | Tomato line or cv. | | | | | |
RAPD marker coupling *Tm-2a*	GCR26 (+)	Ponderosa (+)	GCR267 (*Tm-2a*)	Zuiei (*Tm-2a*)	Paster-Yoozu (*Tm-2a*)	L.p.PI128650 (*Tm-2a*)
$E16_{900}$(GCR236,267)	−	−	+	+	+	+
$N13_{1000}$(GCR236,267)	−	−	+	+	+	+
$G09_{700}$(GCR236,267)	−	−	+	+	+	+
$H09_{1000}$(GCR236,267)	−	−	+	+	+	+
$I15_{2500}$(GCR236,267)	−	−	+	+	+	+
$F02_{2500}$(GCR267)	−	−	+	+	+	+
$H11_{1500}$(GCR267)	−	−	+	+	+	+
$M09_{800}$(GCR267)	−	−	+	+	+	+

(*Dra*I, *Eco*RI, *Eco*RV and *Hin*dIII)-digested total DNA samples prepared from GCR26 (+/+, +/+) and GCR236 (*Tm-2*/*Tm-2*, *nv*/*nv*) (Fig. 4). A clone from OPI18$_{900}$ that is 0.7 cM away from the *nv* and *Tm-2* loci hybridized to one single band in each of the restriction endonuclease-digested DNA samples from GCR236 and most of those from GCR26 (Fig. 4a). In *Dra*I- and *Eco*RI-digested samples, restriction fragment length polymorphisms were observed. A clone from OPB12$_{1200}$ that is also 0.7 cM away from the *nv* and *Tm-2* loci exhibited a hybridization pattern completely different from that of the OPI18$_{900}$ clone (Fig. 4b). This OPB12$_{1200}$ clone hybridized to a considerable numbers of bands in every tomato line. The sequences with homology to this clone appear to be dispersed among chromosomes, but the sequences in some loci may tandemly repeat.

Clones from OPG10$_{800}$, E16$_{900}$ and N13$_{1000}$ hybridized to a number of bands that may be dispersed over the chromosomes (Fig. 4c,d,e). Because some bands exhibited very strong signals, they appear to contain tandem repeats. As mentioned above, we presume that E16$_{900}$ and N13$_{1000}$ are markers nearest to the *nv* and *Tm-2* loci, and thus they may be important as key markers to investigate the structure around the active genes. The clone from OPG09$_{700}$ was shown to be derived from a sequence that is very highly repeated amongst the chromosomes (Fig. 4f).

In this study, we have characterized only six RAPD marker clones, but these have revealed that there are many different kinds of repeated sequences around the active *nv*, *Tm-2* and *Tm-2a* genes. In the next step, physical distances of the markers from the active genes, and genetical

Fig. 4. Southern blots of genomic DNAs from ToMV-resistant and susceptible lines with six RAPD clones (a-f). Each DNA sample was digested with one of the four restriction endonucleases *Dra*I, *Eco*RI, *Eco*RV and *Hin*dIII, separated by 1% agarose gel electrophoresis, blotted on nylon membrane and hybridized with the DIG (digoxigenin)-labeled probes prepared from the clone by using the DIG DNA Labeling System (Boehringer Mannheim).

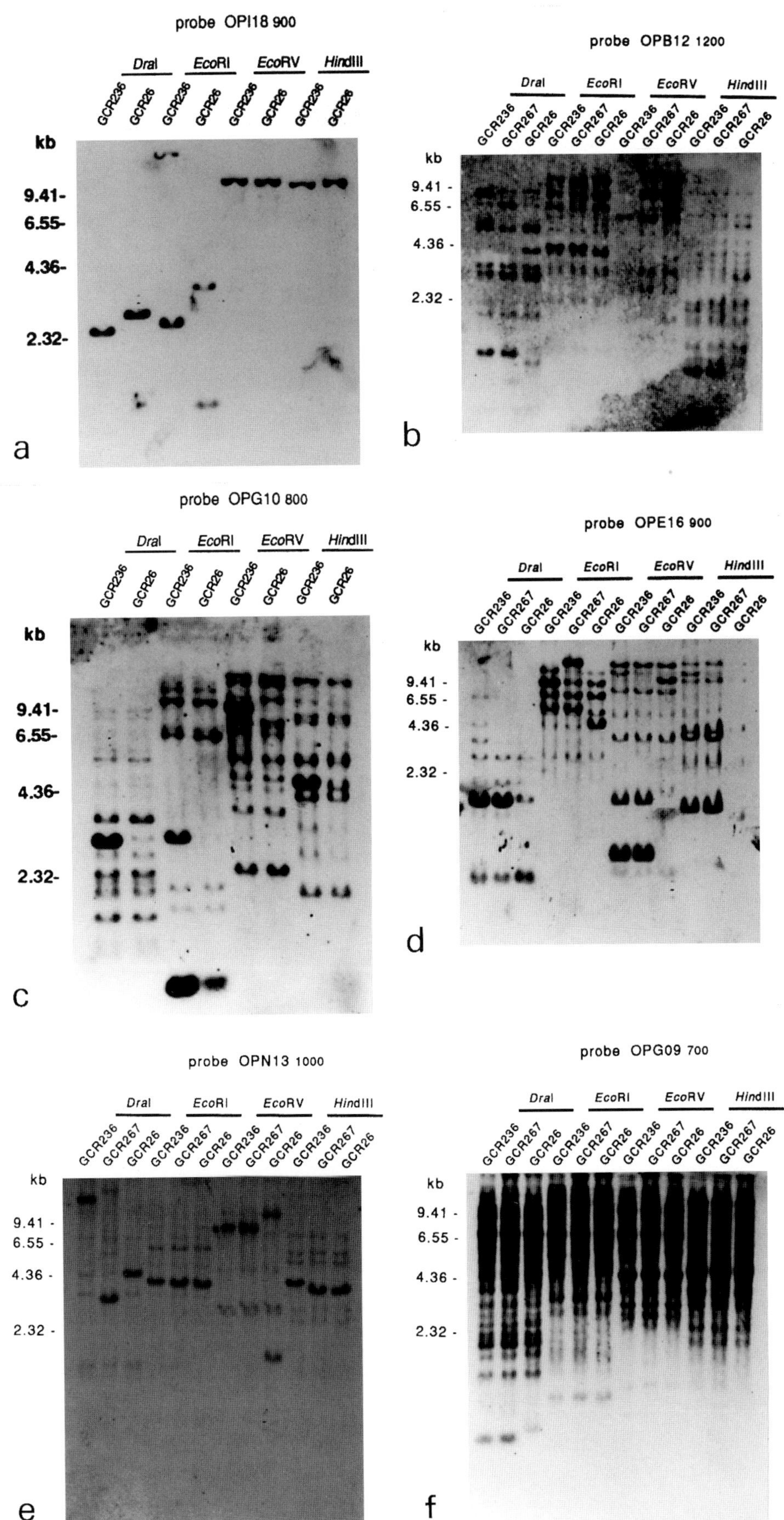

distance in more detail must be investigated using more DNA markers to understand the structures around the genes and their influence on the genetic activity.

REFERENCES

Adam-Blondon, A. F., Sevignac, M., Bannerot, H. and Dron, M. (1994). SCAR, RAPD and RFLP markers linked to a dominant gene (Are) conferring resistance to anthracnose in common bean. *Theor. Appl. Genet.* **88**, 865-870.

Alexander, L. J. (1963). Host-pathogen dynamics of tobacco mosaic virus on tomato. *Phytopathology* **61**, 611-617.

Clayberg, C. D., Butler, L., Rick, C. M. and Young, P. A. (1960). Second list of known genes in the tomato. *J. Hered.* **51**, 167-174.

Ganal, M. W., Lapitan, N. L. V. and Tanksley, S. D. (1988). A molecular and cytogenetic survey of major repeated DNA sequences in tomato (Lycopersicon esculentum). *Mol. Gen. Genet.* **213**, 262-268.

Ganal, M. W., Young, N. D. and Tanksley, S. D. (1989). Pulsed field gel electrophoresis and physical mapping of large DNA fragments in the *Tm-2a* region of chromosome 9 in tomato. *Mol. Gen. Genet.* **215**, 395-400.

Khush, G. S., Rick, C. M. and Robinson, R. W. (1964). Genetic activity in a heterochromatic chromosome segment of the tomato. *Science* **145**, 1432-1434.

Laterrot, H. and Pecaut, P. (1969). Gene Tm2 = new source. *TGC Rep.* **19**, 13-14.

Maisonneuve, B., Bellec, Y., Anderson, P. and Michelmore, R. W. (1994). Rapid mapping of two genes for resistance to downy mildew from Lactuca serriola to existing clusters of resistance genes. *Theor. Appl. Genet.* **89**, 96-104.

Meshi, T., Motoyoshi, F., Maeda, T., Yoshiwoka, S., Watanabe, H. and Okada, Y. (1989). Mutations in the tobacco mosaic virus 30-kDa protein gene overcome *Tm-2* resistance in tomato. *Plant Cell* **1**, 515-522.

Murata, M., Ogura, Y. and Motoyoshi, F. (1994). Centromeric repetitive sequences in *Arabidopsis thaliana. Jpn J. Genet.* **69**, 361-370.

Nishiguchi, M. and Motoyoshi, F. (1987). Resistance mechanisms of tobacco mosaic virus strains in tomato and tobacco. In *Plant Resistance to Viruses* (ed. D. Evered and S. Harnett), pp. 38-46. John Wiley & Sons, Chichester.

Ohmori, T., Murata, M. and Motoyoshi, F. (1995). Identification of RAPD markers linked to the *Tm-2* locus in tomato. *Theor. Appl. Genet.* **90**, 307-311.

Ohno, T., Aoyagi, M., Yamanashi, Y., Saito, H., Ikawa, S., Meshi, T. and Okada, Y. (1984). Nucleotide sequence of the tomato mosaic virus (tomato strain) genome and comparison with the common strain genome. *J. Biochem.* **96**, 1915-1923.

Paran, I. and Michelmore, R. W. (1993). Development of reliable PCR-based markers linked to downy mildew resistance genes in lettuce. *Theor. Appl. Genet.* **85**, 985-993.

Pecaut, P. (1965). Tomate. II. Résistance au virus de la mosaïque du tabac (TMV). *Rep. Station d'Amel. des Plantes Marâichères (INRA)* **1964**, 50-53.

Rick, C. M. (1971). Some cytogenetic features of the genome in diploid plant species. *Stadler Genet. Symp.* **1**, 153-174.

Snoad, B. (1963). Tomato chromosomes and their relationship to the linkage maps. *Proceedings of 11th International Congress of Genetics* **1**, 117.

Smith, J. W. M. and Ritchie, D. B. (1983). A collection of near-isogenic lines of tomato: research tool of the future? *Plant Mol. Biol. Rep.* **1**, 41-45.

Soost, R. K. (1963). Hybrid tomato resistant to tobacco mosaic virus. *J. Hered.* **54**, 241-244.

Yamakawa, K. and Nagata, T. (1975). Three tomato lines obtained by the use of chronic gamma radiation with combined resistance to TMV and Fusarium race J-3. Technical News 16, *Inst. Rad. Breed.*

Printed in Great Britain © The Society for Experimental Biology 1996
SEB0033

The physical organization of Triticeae chromosomes

Trude Schwarzacher

Cereals, John Innes Centre, Colney Lane, Norwich NR4 7UH, UK

(e-mail: Schwarza@bbsrc.ac.uk)

SUMMARY

Molecular cytogenetics combines molecular information of DNA sequences with their chromosomal organization. Genomic in situ hybridization using total genomic DNA as a probe is proving particularly useful to paint chromosomes originating from different genomes in hybrids, alloploid species and alien plant breeding lines. Both the numbers and morphologies of alien chromosomes or chromosome segments can be detected at metaphase and interphase. The method also gives considerable information about species relationships and the distribution of common or diverse DNA sequences between closely related species. Painted chromosomes can be followed through all stages of the cell cycle of somatic and meiotic division, providing new information about chromosome behaviour and pairing at meiosis. In situ hybridization with defined probes enables the physical location of particular DNA sequences to be examined along chromosomes and the analysis of the long range organization of specific chromosome regions. The generation of an integrated genetical, physical and functional map will be useful for the understanding of the organization and structure of the cereal genome.

Key words: In situ hybridization, Recombination, Meiosis, Chromosome painting, Telomere associated DNA sequences

INTRODUCTION

Fluorescence in situ hybridization (FISH) has revolutionized plant and animal cytogenetics and molecular biology. The wide availability of suitable cloned DNA sequences for use as probes, various DNA labelling and other molecular systems, different non-radioactively conjugated nucleotides and different fluorescence detection systems, allowing multi-target probing, have all contributed to the extensive application of the methods. The great potential of in situ hybridization arises from its unique ability to combine cytological information about chromosome and nuclear morphology with molecular information about DNA sequence structure. Hence, the method has been used in mapping projects, and in studies of species relationship and evolution, genome and sequence organization, and nuclear architecture.

In many cases, FISH probes are specific, cloned DNA sequences which include tandemly or dispersed, low copy or repeated, genes or non-coding sequences while an alternate approach uses total genomic DNA (see Heslop-Harrison and Schwarzacher, 1996). In this chapter, I summarize some of our recent work using FISH with total genomic DNA and cloned probes to analyze Triticeae chromosome organization at somatic metaphase and interphase and throughout meiosis.

GENOMIC IN SITU HYBRIDIZATION TO STUDY ALIEN INTROGRESSION AND SPECIES RELATIONSHIPS

In situ hybridization has proved to be very valuable for identifying chromosomes and chromosome segments of different origin (Schwarzacher et al., 1989, 1992). Labelled total genomic DNA from one of the species involved in the hybrid can be used as a probe, and has often been found to be specific enough to show the alien chromosome or chromosome segment. In brief, the method involves labelling total genomic DNA from one species with molecules such as biotin or digoxigenin for use as a probe. Chromosome spreads are made from root tips of the plants. The probe is denatured to make the DNA single stranded, added to the chromosome preparation, and heated to denature the chromosome preparation. The probe and target, chromosomal DNA, are allowed to reaneal to form double-stranded helixes overnight. We then detect sites of stable hybridization using fluorescent labels. Recent protocols for FISH are published (e.g. Heslop-Harrison et al., 1990; Schwarzacher et al., 1994; Anamthawat-Jónsson and Reader, 1995); updates are available from the author, so technical aspects of the methods will not be discussed here.

When labelled genomic DNA from one of the ancestral species in a hybrid is used in combination with a relatively high concentration of blocking DNA from the other species, the discrimination is improved further (Anamthawat-Jónsson et al., 1990; see Fig. 1a and b). Alternatively, each ancestral DNA can be labelled with a different fluorochrome and when visualized together differentiation is greatly enhanced (see Fig. 1c). The ability to use total genomic DNA as a probe to paint plant chromosomes was discovered at the same time as painting methods for mammalian chromosomes were developed (Heslop-Harrison et al., 1988; Pinkel et al., 1988; Lichter et al., 1988). Genomic in situ hybridization permits characterization of the genomes and chromosomes in hybrid plants, allopolyploid species and recombinant breeding lines (Schwarzacher et al., 1989, 1992; Heslop-Harrison et al., 1990; Mukai and Gill,

1991; Bennett et al., 1992; Jiang et al., 1994; Jiang and Gill, 1994; Orgaard et al., 1995). In plant breeding programmes, alien chromosomes can be identified, counted and characterized not only in high-quality metaphase spreads, but also within interphase nuclei, and hence can be easily followed in the initial amphidiploids, the subsequent backcrosses and breeding lines (Heslop-Harrison et al., 1990; Heslop-Harrison and Schwarzacher, 1993). The method also offers new opportunities in phylogenetic and taxonomic studies for determining and testing genomic relationships of wild and cultivated plant species (Heslop-Harrison and Schwarzacher, 1996). Genomic hybridization provides unique data about similarities between DNA from related species, as well as the physical locations of conserved or diverged sequences on chromosomes. Since chromosomes are painted not only at metaphase, but during all stages of the cell cycle, it has the power to confer information about nuclear organization and meiotic chromosome behaviour and pairing configurations.

Fig. 1a shows a somatic root tip metaphase of a wheat line that incorporates both a translocation between wheat and rye (1BL-1RS) and has an additional pair of chromosomes originating from *Thinopyrum bessarabicum*. The preparation has been simultaneously probed with DNA from *Th. bessarabicum* labelled with rhodamine (hybridization sites are detected with red fluorescence), while the rye DNA was labelled with fluorescein (detected by green fluorescence). Fig. 1a is a processed image overlapping the two in situ hybridization colours onto the blue DAPI fluorescence of the chromosomes. The alien chromosomes or chromosome arms are clearly visible in their respective colours and can be counted, analysed and measured, while very little cross hybridization of the alien DNA probes is detectable on wheat chromatin. Alien chromosomes are also visible in interphase nuclei (Fig. 1b) and can be counted quickly and easily in breeding lines where the availability of complete and well spread metaphase plates can be restricted.

In Fig. 1a, it is notable that approximately fourteen smaller wheat chromosomes, probably of D genome origin, show more cross hybridization with the red *Th. bessarabicum* probe indicating that *Th. bessarabicum* is more closely related to the D genome than the A and B genomes of wheat (see also Heslop-Harrison and Schwarzacher (1996). Genomic in situ hybridization is also able to detect chromosome regions that contain either mainly species specific DNA sequences or wide spread sequences. For example, genomic in situ hybridization was used to show regions of *Hordeum* chromosomes which differ between *H. vulgare* and *H. bulbosum* (Anamthawat-Jónsson et al., 1993) or the differences and similarities between the related *Leymus* and *Psathyrostachys* genera (Orgaard and Heslop-Harrison, 1994a,b). In these cases, the alien, but related, genome shows positive bands containing sequences that are in common between it and the species from which the probe is derived, while there are negative bands containing essential species-specific sequences.

Genetic relationships between species have often been established by the extent to which the chromosomes are paired at metaphase I of meiosis (see Dewey, 1984; Jauhar, 1996). However, pairing at pachytene of meiotic prophase can be far more extensive than indicated by metaphase I configurations. For example, in hexaploid wheat, many pairing partner switches were found at pachytene while only bivalents are present almost exclusively at metaphase I, and multivalents in

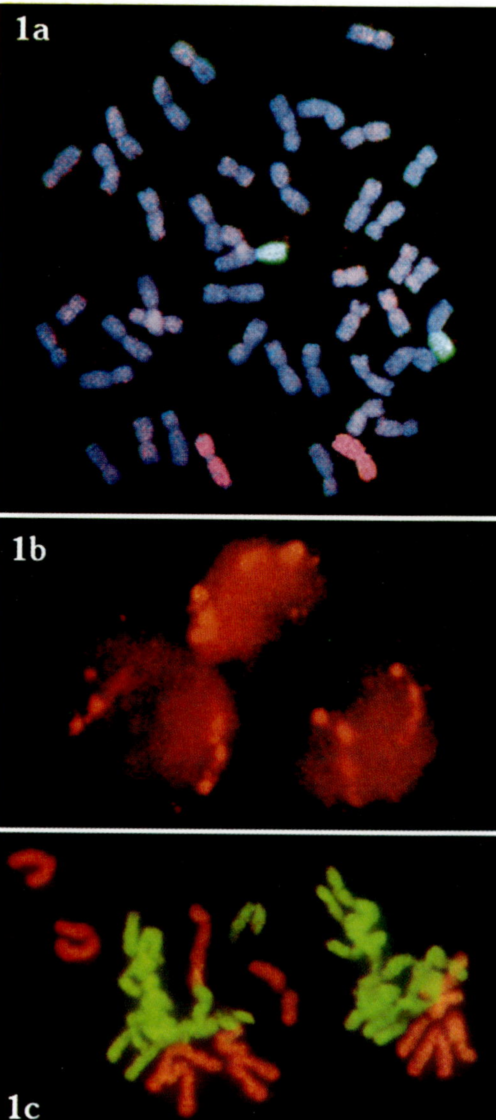

Fig. 1. Examples of genomic in situ hybridization to identify the origin of chromatin in alien-wheat recombinant lines (a,b) and the hybrid *Hordeum procerum* × *Leymus racemosus* (c). (a) A processed overlay picture of a metaphase (2n=44) from a wheat line containing a rye wheat translocation (1BL/1RS) and an added pair of *Th. bessarabicum* chromosomes after double target FISH and staining with DAPI. Total genomic DNA from rye labelled with fluorescein identifies the rye chromosome arms by green fluoresence and rhodamine labelled DNA from *Th. bessarabicum* identifies the *Th. bessarabicum* chromosomes by red fluorescence, while the wheat chromosomes show the blue fluroescence of the DAPI counterstain. At interphase (b) the two rye chromosome arms are separated, but show concerted coiling patterns. (c) Anaphase chromosomes in the hybrid *H. procerum* × *L. racemosus* were identified by genomic in situ hybridization using Texas red and fluorescein which were visualized together through a double band pass filter set.

tetraploid *Scilla*s are eliminated during meiotic prophase (see Jenkins and Rees, 1991). Genomic in situ hybridization allows not only the painting of individual condensed chromosomes at somatic division (Fig. 1), but useful analyses of whole mounts of chromosomes at all the stages of meiosis become possible.

Using genomic in situ hybridization to intergeneric hybrids, it is possible to determine the genomic origin of paired and unpaired chromosome axes at pachytene and of metaphase I configurations. In the hybrid *Hordeum procerum* × *Leymus racemosus*, genomic probing differentiates clearly the two parental genomes, and shows each forming separate domains (Fig. 1c). At pachytene, pairing was observed between chromosome segments of *H. procerum* origin, and *L. racemosus* origin and between the two parents while at metaphase I many univalents and only bivalents of single parental origin were found (Schwarzacher and Heslop-Harrison, 1995; T. Schwarzacher and M. Orgaard, unpublished). Consequently, at anaphase I chromosomes of either parental origin were lagging and formed micronuclei at the diad stage and also at the following tetrad stage. It is likely that analysis at both meiotic prophase and metaphase I will give the maximum amount of information about genetic relationships between homologous and homoeologous chromosomes in a single organism. These data also indicate that pairing and synapsis of chromosome axes and recombination followed by chiasma formation might require different levels of homologies and might be regulated by different mechanisms.

As mentioned above, use of genomic DNA as a probe enables individual chromosome arms to be followed within interphase nuclei. In wheat, we have investigated the location of chromosomes or chromosome arms introduced from alien species (Heslop-Harrison et al., 1991; Schwarzacher et al., 1992). The individual alien cereal chromosome or chromosome arms tend to occupy discrete, elongated domains at interphase. Fig. 1b shows a wheat line with the 1B-1R translocation, and the two rye chromosome domains are visible. Pairs of homologous chromosome arms may show similar patterns of coiling or condensation, although there is no evidence for universal association of homologues: homologous chromosomes show such concerted patterns regardless of their physical separation in the nucleus. The condensation is probably a property of the nuclear structure.

REPEATED DNA SEQUENCES AT THE TELOMERES OF RYE CHROMOSOMES

The regions around the centromeres and telomeres of chromosomes from many species consist of highly repetitive DNA (Harrison and Heslop-Harrison, 1995; Martinez-Zapater et al., 1986; Richards et al., 1993; and see Heslop-Harrison, 1996, this volume). Repetitive DNA sequences in the terminal heterochromatin of rye (*Secale cereale*) chromosomes have consequences for the structural and functional organization of chromosomes. We studied the large-scale genomic organization of the subtelomeric regions using clones of three non-homologous, tandemly repeated, subtelomeric DNA families (pSc119.2, pSc200 and pSc250) that have complex but contrasting higher order structural organization (Vershinin et al., 1995). After polymerase chain reaction analysis with single or pairs of primer(s), apart from the commonly found head-to-tail orientation, a fraction of the repeat units of one family, pSc200, was organized in a head-to-head orientation. Such head-to-head structures are the molecular correspondent of chromosome evolution by chromatid-type breakage-fusion-bridge cycles (McClintock, 1941). After *Xba*I digestion and pulse field gel electrophoresis, the subtelomeric clones showed strong hybridization signals from 40-100 kb, with a maximum at 50 to 60 kb. Vershinin et al. (1995) suggested that these fragments define a basic higher order structure and DNA loop domains of regions of rye chromosomes consisting of arrays of tandemly organized sequences.

Using in situ hybridization, all three repeat families are located in the major subtelomeric heterochromatic regions with pSc200 being most terminal and pSc119.2 most proximal (Fig. 2a and b). pSc200 and pSc250 have a similar hybridization pattern suggesting co-evolution and co-distribution throughout the genome. In contrast, pSc119.2 shows different strengths of hybridization at the terminal bands and also has many intercalary sites, in agreement with its clearly different organization after hybridization to pulse field gel separated DNA digests.

COMPARING PHYSICAL AND GENETIC MAPS; CHIASMATA DISTRIBUTION

Comparisons of genetic and physical maps of large cereal genomes have revealed substantial discrepancies (Heslop-Harrison, 1991; Leitch and Heslop-Harrison, 1993; Lukaszewski, 1992; Werner et al., 1992). Many genes and markers are clustered genetically near the centromere while they are located in the distal region of the chromosome map. Hence, the distal segments include most of the genetic length and recombination of the chromosomes which agrees with the distal location of chiasmata. Distribution of chiasmata and their relation to cross-overs and recombination as scored when constructing genetic maps from linkage data is an area of current interest and controversy (Nilsson et al., 1993; Tease and Jones, 1995). In situ hybridization with an rDNA probe identifies the bivalent of chromosome 1R of rye with the physical rDNA gene location two-thirds down the chromosome arm (Fig. 2c-f). The location of the chiasma, either distal (Fig. 2c and f) or proximal (Fig. 2d) of the rDNA results in significant bivalent appearance and could be scored easily (Table 1). At metaphase I of meiosis, chiasmata were found on the long arm and on the short arm; 84% of chiasmata were distal to the rDNA locus and 16% proximal, corresponding closely to the location on the genetic map of Devos et al. (1993). In situ hybridization with the repeated DNA sequences pSc200 and pSc250 revealed chiasmata exclusively near the subtelomeric heterochromatin of rye bivalents at metaphase-I (Fig. 2b). Sometimes chiasmata were found within the more proximal pSc250 sequences, but not the distal pSc200 sequences, suggesting, in agreement with other analyses, that pSc250 sequences have been amplified mainly via unequal crossing over while pSc200 family evolution also involves breakage-fusion-bridge cycles (Vershinin et al., 1995).

CONCLUSIONS

Genomic in situ hybridization has many prospects for analysis of chromosome behaviour at metaphase, interphase and during meiosis and to understand species relationships (Heslop-Harrison and Schwarzacher, 1996). In situ hybridization using repetitive DNA sequences allows the analysis of the long range

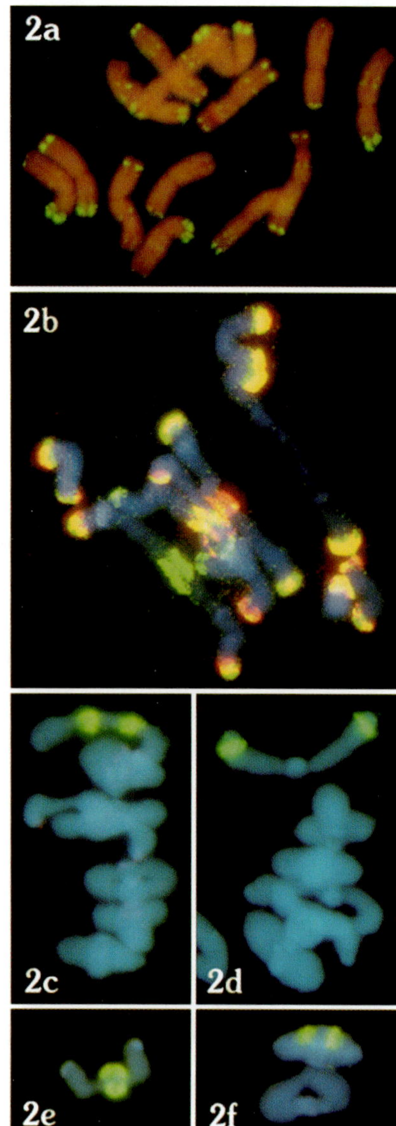

Fig. 2. Rye (*Secale cereale*) chromosomes after in situ hybridization using repetitive DNA probes. (a) A somatic metaphase probed with the clone pSc119.2 (McIntyre et al., 1990); major subtelomeric and a few minor intercalary sites are visble by fluorescein fluorescence (green) on the propidium iodide counterstained chromosomes (red). (b) Double target FISH using pSc200 (green fluorescence) and pSc250 (red fluorescence) to meiotic anaphase I chromosomes showing major sites of both probes at most telomeres (overlapping green and red fluorescence results in yellow). (c-f) Examples of metaphase I figurations of chromosome 1R identified by the green fluorescence of the rDNA probe (pTa71; Gerlach and Bedbrook, 1979; see Table 1); (c) rod bivalent with chiasma on the short arm distal to the NOR; (d) rod bivalent with chiasma on the long arm; (e) rod bivalent with chiasma on the short arm proximal to the NOR, not part of the scoring population; (f) ring bivalent with chiasma on the long arm and chiasma on the short arm distally to the NOR.

organization of the genome and helps to link molecular and sequence data with information about chromosome structure and function (see Vershinin et al., 1995). As in so many areas of genome analysis and cytogenetics, methods of molecular cytogenetics are giving new insight into the process, and

Table 1. Chiasmata distribution at meiotic metaphase I of rye chromosome 1R

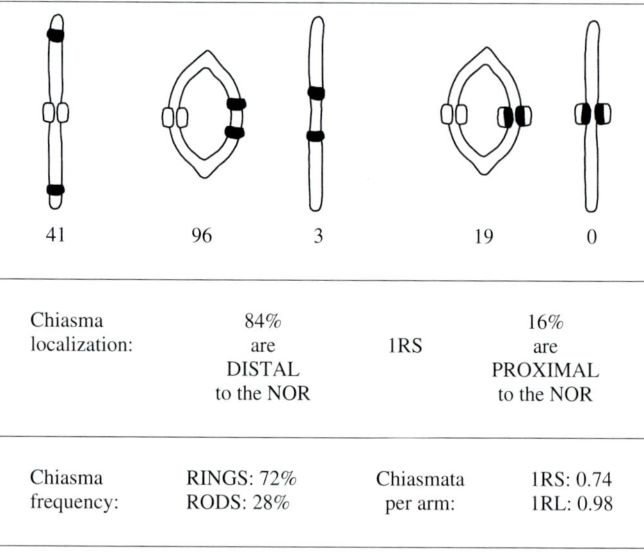

| | 41 | 96 | 3 | 19 | 0 |

Chiasma localization:	84% are DISTAL to the NOR		1RS		16% are PROXIMAL to the NOR
Chiasma frequency:	RINGS: 72% RODS: 28%		Chiasmata per arm:		1RS: 0.74 1RL: 0.98

helping form an integrated genetical, functional, structural and physical map of the genome.

Key questions about many of the fundamental processes and timing of meiosis are still unanswered. In particular, to find when and where chromosomes come together for recombination and meiosis and what DNA sequences are concentrated at sites of recombination and pairing (and their molecular interaction with the synaptonemal complex) are important research objectives. Molecular cytogenetic methods are beginning to enable us to examine when and where chromosome recognition, pairing and recombination occur and may provide answers.

I thank Dr A. V. Vershinin and Professor J. S. Heslop-Harrison for the fruitful collaboration on the organization of the telomeric regions in rye, Mrs Shaobin Wu for her valuable technical assistance, and Dr C. B. Gillies for providing Fig. 2b and many discussions. BBSRC is acknowledged for financial support.

REFERENCES

Anamthawat-Jónsson, K., Schwarzacher, T., Leitch, A. R., Bennett, M. D. and Heslop-Harrison, J. S. (1990). Discrimination between closely related Triticeae species using genomic DNA as a probe. *Theor. Appl. Genet.* **79**, 721-728.

Anamthawat-Jónsson, K., Schwarzacher, T. and Heslop-Harrison, J. S. (1993). Behavior of parental genomes in the hybrid *Hordeum vulgare* × *H. bulbosum*. *J. Hered.* **84**, 78-82.

Anamthawat-Jónsson, K. and Reader, S. M. (1995). Pre-annealing of total genomic DNA probes for simultaneous genomic in situ hybridization. *Genome* **38**, 814-816.

Bennett, S. T., Kenton, A. Y. and Bennett, M. D. (1992). Genomic in situ hybridization reveals the allopolyploid nature of Milium montianum (Gramineae). *Chromosoma* **101**, 420-424.

Devos, K. M., Atkinson, M. D., Chinoy, C. N., Koebner, R. M. D., Francis, H. A., Harcourt, R. L., Liu, C. J., Masojc, P., Xie, D. X. and Gale, M. D. (1993). Chromosomal rearrangements in the rye genome relative to that of wheat. *Theor. Appl. Genet.* **85**, 673-680.

Dewey, D. R. (1984). The genomic system of classification as a guide to intergeneric hybridization with the perennial Triticeae. *Gene Manip. Plant Improve.* 209-279.

Gerlach, W. L. and Bedbrook, J. R. (1979). Cloning and characterization of ribosomal RNA genes from wheat and barley. *Nucl. Acids Res.* **7**, 1869-1885.

Harrison, G. E. and Heslop-Harrison, J. S. (1995). Centromeric repetitive DNA in the genus *Brassica*. *Theor. Appl. Genet.* **90**, 157-165.

Heslop-Harrison, J. S., Schwarzacher, T., Leitch, A. R., Anamthawat-Jónsson, K. and Bennett, M. D. (1988). A method of identifying DNA sequences in chromosomes of plants. European Patent Application Number 8828130.8. December 2.

Heslop-Harrison, J. S., Leitch, A. R., Schwarzacher, T. and Anamthawat-Jónsson, K. (1990). Detection and characterization of 1B/1R translocations in hexaploid wheat. *Heredity* **65**, 385-392.

Heslop-Harrison, J. S. (1991). The molecular cytogenetics of plants. *J. Cell Sci.* **100**, 15-21.

Heslop-Harrison, J. S., Schwarzacher, T., Anamthawat-Jónsson, K., Leitch, A. R., Shi, M. and Leitch, I. J. (1991). *In situ* hybridization with automated chromosome denaturation. *Technique* **3**, 109-115

Heslop-Harrison, J. S. and Schwarzacher, T. (1993). Molecular cytogenetics – biology and applications in plant breeding. *Chrom. Today* **11**, 191-198.

Heslop-Harrison, J. S. (1996). Comparative analysis of plant genome architecture. *SEB Symp.* **50**, 17-23.

Heslop-Harrison, J. S. and Schwarzacher, T. (1996). Genomic Southern and *in situ* hybridization for plant genome analysis. In *Methods of Genome Analysis in Plants* (ed. P. P. Jauhar), pp. 163-179. Boca Raton: CRC.

Jauhar, P. P., editor (1996). *Methods of Genome Analysis in Plants*. Boca Raton: CRC.

Jenkins, G. and Rees, H. (1991). Strategies of bivalent formation in allopolyploid plants. *Proc. R. Soc. Lond. Ser. B Biol. Sci.* **243**, 209-214.

Jiang, J. and Gill, B. S. (1994). Nonisotopic in situ hybridization and plant genome mapping: the first 10 years. *Genome* **37**, 717-725.

Jiang, J., Morris, K. L. D. and Gill, B. S. (1994). Introgression of Elymus trachycaulus chromatin into common wheat. *Chromosome Res.* **2**, 3-13.

Leitch, I. J. and Heslop-Harrison, J. S. (1993). Physical mapping of four sites of 5S rDNA sequences and one site of the alpha-amylase-2 gene in barley (*Hordeum vulgare*). *Genome* **36**, 517-523.

Lichter, P., Cremer, T., Borden, J., Manuelidis, L. and Ward, D. C. (1988). Delineation of individual human chromosomes in metaphase and interphase cells by in situ suppression hybridization using recombinant DNA libraries. *Hum. Genet.* **80**, 224-234.

Lukaszewski, A. J. (1992). A comparison of physical distribution of recombination in chromosome 1R in diploid rye and in hexaploid triticale. *Theor. Appl. Genet.* **83**, 1048-1053.

Martinez-Zapater, J. M., Estelle, M. A. and Somerville, C. R. (1986). A highly repeated DNA sequence in Arabidopsis thaliana. *Mol. Gen. Genet.* **204**, 417-423.

McClintock, B. (1941). The stability of broken ends of chromosomes in *Zea mays*. *Genetics* **26**, 279-282.

McIntyre, C. L., Pereira, S., Moran, L. B. and Appels, R. (1990). New Secale cereale (rye) DNA derivatives for the detection of rye chromosome segments in wheat. *Genome* **33**, 635-640.

Mukai, Y. and Gill, B. S. (1991). Detection of barley chromatin added to wheat by genomic *in situ* hybridization. *Genome* **34**, 448-452.

Nilsson, N.-O., Sall, T. and Bengtsson, B. O. (1993). Chiasma and recombination data in plants – are they compatible? *Trends Genet.*

Orgaard, M. and Heslop-Harrison, J. S. (1994a). Relationships between species of *Leymus, Psathyrostachys* and *Hordeum* (Poaceae, Triticeae) inferred from Southern hybridization of genomic DNA and cloned DNA probes. *Plant Syst. Evol.* **189**, 217-231.

Orgaard, M. and Heslop-Harrison, J. S. (1994b). Investigations of genome relationships in *Leymus, Psathyrostachys* and *Hordeum* by genomic DNA:DNA *in situ* hybridization. *Ann. Bot.* **73**, 195-203.

Orgaard, M., Jacobsen, N. and Heslop-Harrison, J. S. (1995). The hybrid origin of two cultivars of *Crocus* (Iridaceae) analysed by molecular cytogenetics including genomic Southern and *in situ* hybridization. *Ann. Bot.* **76**, 253-262.

Pinkel, D., Landegent, J., Collins, C., Fuscoe, J., Segraves, R., Lucas, J. and Gray, J. (1988). Fluorescence in situ hybridization with human chromosome-specific libraries: detection of trisomy 21 and translocations of chromosome 4. *Proc. Nat. Acad. Sci. USA* **85**, 9138-9142.

Richards, E. J., Vongs, A., Walsh, M., Yang, J. and Chao, S. (1993). Substructure of telomere repeat arrays. In *The Chromosome* (ed. J. S. Heslop-Harrison and R. B. Flavell), pp. 103-114. Oxford: BIOS.

Schwarzacher, T., Leitch, A. R., Bennett, M. D. and Heslop-Harrison, J. S. (1989). *In situ* localization of parental genomes in a wide hybrid. *Ann. Bot.* **64**, 315-324.

Schwarzacher, T., Anamthawat-Jónsson, K., Harrison, G. E., Islam, A. K. M. R., Jia, J. Z., King, I. P., Leitch, A. R., Miller, T. E., Reader, S. M., Rogers, W. J., Shi, M. and Heslop-Harrison, J. S. (1992). Genomic *in situ* hybridization to identify alien chromosomes and chromosome segments in wheat. *Theor. Appl. Genet.* **84**, 778-786.

Schwarzacher T., Leitch A. R. and Heslop-Harrison J. S. (1994). DNA:DNA *in situ* hybridization – methods for light microscopy. In *Plant Cell Biology: A Practical Approach* (ed. N. Harris and K. J. Oparka), pp. 127-155. Oxford: Oxford University Press.

Schwarzacher, T. and Heslop-Harrison, J. S. (1995). Molecular cytogenetic investigations of meiosis. In *Kew Chromosome Conference IV* (ed. P. Brandham and M. D. Bennett), pp. 407-416+xl. Kew, London, UK: Royal Botanic Gardens.

Tease, C. and Jones, G. H. (1995). Do chisamata disappear? An examination of whether closely spaced chiasmata are liable to reduction or loss. *Chromosome Res.* **3**, 162-168.

Vershinin, A., Schwarzacher, T. and Heslop-Harrison, J. S. (1995). The large scale genomic organization of repetitive DNA families at the telomeres of rye chromosomes. *Plant Cell* **7**, 1823-1833.

Werner, J. E., Endo, T. R. and Gill, B. S. (1992). Toward a cytogenetically based physical map of the wheat genome. *Proc. Nat. Acad. Sci. USA* **89**, 11307-11311.

Subject Index

Abies, 'experimental taxonomy': JACOBSEN AND ØRGAARD 61
Ancestral genome: BENNETT 45
Arabidopsis genome, sequencing and mapping: DELSENY, RAYNAL, LAUDIE, VAROQUAUX, COMELLA, WU, COOKE AND GRELLET 5

Chromosomes
   high-resolution mapping: HERRMANN, MARTIN, BUSCH, WANNER AND HOHMANN 25
   painting, *Triticeae*: SCHWARZACHER 71
   ultrastucture: HERRMANN, MARTIN, BUSCH, WANNER AND HOHMANN 25
   walking: BENNETZEN, SANMIGUEL, LIU, CHEN, TIKHONOV, COSTA DE OLIVEIRA, JIN, AVRAMOVA, WOO, ZHANG AND WING 1
Collinearity, in evolution of grass nuclear genomes: BENNETZEN, SANMIGUEL, LIU, CHEN, TIKHONOV, COSTA DE OLIVEIRA, JIN, AVRAMOVA, WOO, ZHANG AND WING 1
Comparative genome analysis
   HERRMANN, MARTIN, BUSCH, WANNER AND HOHMANN 25
   HESLOP-HARRISON 17
Crocus, 'experimental taxonomy': JACOBSEN AND ØRGAARD 61

DNA
   cDNA, rice, as model for expressed genes: SASAKI 11
   rDNA, slash pine: DOUDRICK 53
   repetitive: HESLOP-HARRISON 17
Drought tolerance, in tropical maize, QTL for: HOISINGTON, JIANG, KHAIRALLAH, RIBAUT, BOHN, WILLCOX AND GONZALEZ-DE-LEON 39

Evolution, and 'experimental taxonomy': JACOBSEN AND ØRGAARD 61
Expressed sequence tags: DELSENY, RAYNAL, LAUDIE, VAROQUAUX, COMELLA, WU, COOKE AND GRELLET 5

FISH (fluorescence in situ hybridization): MOTOYOSHI, OHMORI AND MURATA 65

Genetic mapping
   slash pine: DOUDRICK 53
   sorghum and maize: LEE 31
Genetic recombinational analysis, slash pine: DOUDRICK 53
Genome, ancestral: BENNETT 45
Genome analysis, comparative *see* Comparative genome analysis
Genome mapping
   Arabidopsis: DELSENY, RAYNAL, LAUDIE, VAROQUAUX, COMELLA, WU, COOKE AND GRELLET 5
   high-resolution: HERRMANN, MARTIN, BUSCH, WANNER AND HOHMANN 25
Genome organization: HESLOP-HARRISON 17
   in grass: BENNETZEN, SANMIGUEL, LIU, CHEN, TIKHONOV, COSTA DE OLIVEIRA, JIN, AVRAMOVA, WOO, ZHANG AND WING 1
Genome sequencing, *Arabidopsis*, DELSENY, RAYNAL, LAUDIE, VAROQUAUX, COMELLA, WU, COOKE AND GRELLET 5
Grass, microcollinearity and segmental duplication: BENNETZEN, SANMIGUEL, LIU, CHEN, TIKHONOV, COSTA DE OLIVEIRA, JIN, AVRAMOVA, WOO, ZHANG AND WING 1

Heterochromatin, *Tm-2* locus in tomato: MOTOYOSHI, OHMORI AND MURATA 65
Hordeum, 'experimental taxonomy': JACOBSEN AND ØRGAARD 61

In situ hybridization: HESLOP-HARRISON 17
   slash pine: DOUDRICK 53
   Triticeae: SCHWARZACHER 71

Insect resistance, in tropical maize, QTL for: HOISINGTON, JIANG, KHAIRALLAH, RIBAUT, BOHN, WILLCOX AND GONZALEZ-DE-LEON 39

Karyotype
   natural: BENNETT 45
   slash pine: DOUDRICK 53
Lilium, 'experimental taxonomy': JACOBSEN AND ØRGAARD 61

Linkage analyses, slash pine: DOUDRICK 53

Maize
   microcollinearity and segmental duplication: BENNETZEN, SANMIGUEL, LIU, CHEN, TIKHONOV, COSTA DE OLIVEIRA, JIN, AVRAMOVA, WOO, ZHANG AND WING 1
   QTL for insect resistance and drought tolerance: HOISINGTON, JIANG, KHAIRALLAH, RIBAUT, BOHN, WILLCOX AND GONZALEZ-DE-LEON 39
   QTL mapping: LEE 31
Mapping *see* Genome mapping
Meiosis, in *Triticeae*: SCHWARZACHER 71
Microcollinearity, in evolution of grass nuclear genomes: BENNETZEN, SANMIGUEL, LIU, CHEN, TIKHONOV, COSTA DE OLIVEIRA, JIN, AVRAMOVA, WOO, ZHANG AND WING 1

Norway spruce (*Picea abies*), 'experimental taxonomy': JACOBSEN AND ØRGAARD 61
Nuclear architecture: HESLOP-HARRISON 17
Nucleotype: BENNETT 45

Pachyrhizus, 'experimental taxonomy': JACOBSEN AND ØRGAARD 61
Picea abies (Norway spruce), 'experimental taxonomy': JACOBSEN AND ØRGAARD 61
Pinus elliotti (slash pine), linkage analyses: DOUDRICK 53
Plant molecular data: JACOBSEN AND ØRGAARD 61

QTL (quantitative trait loci)
   for insect resistance and drought tolerance in tropical maize: HOISINGTON, JIANG, KHAIRALLAH, RIBAUT, BOHN, WILLCOX AND GONZALEZ-DE-LEON 39
   mapping, in sorghum and maize: LEE 31
   RAPD (random amplified polymorphic DNA), *Tm-2* locus in tomato: MOTOYOSHI, OHMORI AND MURATA 65

Recombination
   slash pine: DOUDRICK 53
   Triticeae: SCHWARZACHER 71
RFLPs (restriction fragment length polymorphisms): SASAKI 11
   in insect resistance and drought tolerance in tropical maize:HOISINGTON, JIANG, KHAIRALLAH, RIBAUT, BOHN, WILLCOX AND GONZALEZ-DE-LEON 39
   in sorghum and maize: LEE 31
Rice
   cDNA as model for expressed genes: SASAKI 11
   microcollinearity and segmental duplication: BENNETZEN, SANMIGUEL, LIU, CHEN, TIKHONOV, COSTA DE OLIVEIRA, JIN, AVRAMOVA, WOO, ZHANG AND WING 1

SCAR (sequence characterized amplified region) markers, *Tm-2* locus in tomato: MOTOYOSHI, OHMORI AND MURATA 65
Similarity search: SASAKI 11
Slash pine (*Pinus elliotti*), linkage analyses: DOUDRICK 53
Sorghum
   microcollinearity and segmental duplication: BENNETZEN, SANMIGUEL, LIU, CHEN, TIKHONOV, COSTA DE OLIVEIRA, JIN, AVRAMOVA, WOO, ZHANG AND WING 1

QTL mapping: LEE 31
Synteny: HERRMANN, MARTIN, BUSCH, WANNER AND
    HOHMANN 25

Tandem duplication: BENNETZEN, SANMIGUEL, LIU, CHEN,
    TIKHONOV, COSTA DE OLIVEIRA, JIN, AVRAMOVA, WOO,
    ZHANG AND WING 1
Telomere
    in slash pine: DOUDRICK 53
    in *Triticeae*, associated DNA sequence: SCHWARZACHER 71

Tm-2 locus in tomato: MOTOYOSHI, OHMORI AND MURATA
    65
Tomato, *TM-2* locus: MOTOYOSHI, OHMORI AND MURATA 65
Tomato mosaic virus (ToMV): MOTOYOSHI, OHMORI AND
    MURATA 65
Triploid pathway: JACOBSEN AND ØRGAARD 61
Triticeae chromosomes
    high resolution mapping: HERRMANN, MARTIN, BUSCH,
        WANNER AND HOHMANN 25
    physical organization of: SCHWARZACHER 71